Economic and Societal Impacts of

TORNADOES

KEVIN M. SIMMONS AND DANIEL SUTTER

AMERICAN METEOROLOGICAL SOCIETY

Front cover photograph © University Corporation for Atmospheric Research, photo by Robert Henson.

All photos courtesy Texas Tech University unless otherwise noted. The authors wish to thank Texas Tech University's Wind Science and Engineering (WISE) Research Center and their Wind Damage Documentation Library for the use of photos throughout the book. Many thanks to WISE Ph.D. students Ian Giammanco and Patrick Skinner for their storm chase photos; many thanks to WISE Ph.D. students Richard Krupar and Kelsey Seger for help sorting through 40 years' worth of tornado damage photographs.

Published by the American Meteorological Society
45 Beacon Street, Boston, Massachusetts 02108

The mission of the American Meteorological Society is to advance the atmospheric and related sciences, technologies, applications, and services for the benefit of society. Founded in 1919, the AMS has a membership of more than 13,000 and represents the premier scientific and professional society serving the atmospheric and related sciences. Additional information regarding society activities and membership can be found at www.ametsoc.org.

For more AMS Books, see www.ametsoc.org/amsbookstore. Order online, or call (617) 226-3998.

Library of Congress Cataloging-in-Publication Data

Simmons , Kevin M.
  Economic and societal impacts of tornadoes / Kevin M. Simmons and Daniel Sutter.
    p. cm.
  ISBN 978-1-878220-99-8 (pbk.)
  1. Tornadoes—Economic aspects—United States. 2. Tornadoes—Social aspects—United States. 3. Mass casualities—Prevention—United States. 4. Emergency management—United States. I. Sutter, Daniel. II. Title.
  HV635.5.S486 2011
  363.34'92340973—dc22
                    2010051906

Mixed Sources
Product group from well-managed forests, controlled sources and recycled wood or fiber
www.fsc.org Cert no. SW-COC-003925
© 1996 Forest Stewardship Council
FSC

CONTENTS

DEDICATION

Kevin
To Susan, who is still as lovely to me today as when we attended our high school prom together.

Dan
To my parents, who always thought that professors should be writing books.

Special Thanks

We would like to thank NOAA's National Severe Storms Laboratory and the Institute for Catastrophic Loss Reduction for funding provided throughout our work on this research agenda. We also give a special thanks to Jeff Kimpel, Bill Hooke, Harold Brooks, Jamie Kruse, Paul Kovacs, and Kishor Mehta for encouraging our work on this topic. The project would not have been possible without the generous sharing of data from Joe Schaefer, Brent Macaloney, and Tim Crum. Finally, Dan wants to thank Natalie for her support and encouragement of his research, and his dogs Norm and Cliff for their patience in having their walks and playtime disrupted during his work on the project.

FOREWORD

The economics of natural disasters is an area of rising importance for the economics profession and also for the world more generally. For decades, economists considered this area to stand outside the normal interests of the science. But in these days of global warming, the earthquake in Haiti, flooding in Pakistan, Hurricane Katrina, and rampaging forest fires in Russia, it has become clear that natural disasters are at the very center of the problem of economic and social development.

In fact, economic disasters bring together many of the central features of economics. Disasters, including tornadoes, test the flexibility and resilience of economic and political institutions. They pose the question of how an area or a neighborhood will recover, how it will recover its ability to mobilize resources, and how it will move from a situation of lesser wealth to greater wealth. It is, sadly, a perfect controlled natural experiment to see both how wealth is destroyed and how wealth is created again.

Tornadoes are also a proxy for the larger idea that economic development is not a smooth process. Economic development involves some large and discrete steps backwards, often followed by some significant leaps forward. Tornadoes, and the recovery from tornadoes, show these same processes.

Natural disasters are important for other reasons as well. How we prepare for these disasters indicates our tolerance for risk and our ability to insure and self-insure. It shows in which ways we value human lives, and in which ways we are willing to lose some lives to minimize protective expenditures. How we treat and compensate victims reflects our sense of justice and fairness. How we rebuild demonstrates our sense of the future and how much optimism we will throw at solving a problem. It reflects some fundamental truths about how politics works and, sometimes, how politics fails.

A study of natural disasters should be of interest to all economists. It is also the case that this area is badly understudied and thus there is much low-hanging fruit to be had. It is this low-hanging fruit that you will find in the new book by Kevin M. Simmons and Daniel Sutter.

Simmons and Sutter already have staked out their ground in this area with numerous academic journal publications over the last ten years. It is now time for their work to be turned into comprehensive book form. Each year, about 1,200 tornadoes touch down across the United States, but to date we have not seen a book of this detail and analytical fortitude.

It is appropriate that this book is sponsored by the American Meteorological Society and distributed by the University of Chicago Press. This combination represents the integration of theory and practice that the authors develop so successfully.

Tyler Cowen
Professor of Economics,
George Mason University

1

WHAT WE CAN LEARN FROM SOCIETAL IMPACTS ANALYSIS

What does supply and demand teach us about whether rotating wall clouds will spin out a tornado? Nothing, really. Can we use the stock market to understand why tornadoes can be so capricious, flattening one house and leaving the one next door untouched? Well, no. So why would a couple of economists who have never even seen a tornado except on TV, and who know nothing about cloud dynamics, write a book about tornadoes? And why would anyone want to read it?

In fact, we have not written a book about tornadoes. Our subject is the economic and societal impact of tornadoes. As the geographer Gilbert White taught generations of students and scholars, societal impacts arise from the interaction of nature and humans. Tornadoes are natural events, but a tornado disaster has a human component.

1.1. Our Approach

This book is economic in its methods: We apply models and statistical methods from economics. Readers with a narrow view of economics—who think of economic topics as having to do exclusively with money and business—

may find the impacts we examine not to be economic at all, and will probably characterize this book as about societal impacts only.

Our primary focus will be on casualties, not property damage or business impacts. We seek to analyze and understand the various impacts of tornadoes on society. We also analyze the effects of efforts to reduce impacts, such as tornado warnings and watches, and tornado shelters and safe rooms. Our analysis is primarily positive, in that we seek to identify patterns and evaluate mitigative efforts. But our positive analysis is conducted with an eye toward eventually being able to offer suggestions about how impacts, and again primarily casualties, can be reduced in a cost-effective manner.

1.2. Research, Tornadoes, and Societal Impacts: A Short History

Atmospheric scientists have learned a great deal about tornadoes in recent decades, much of it through several large research projects that were cooperative ventures between the government and leading universities. One of the more famous projects is VORTEX (Verification of the Origins of Rotation in Tornadoes Experiment), conducted in the spring seasons of 1994 and 1995. The project was led by Dr. Eric Rasmussen and coordinated by the National Severe Storms Laboratory in Norman, Oklahoma, and included as partners the University of Oklahoma, Texas A&M University, the University of Illinois, Texas Tech University, New Mexico Tech, West Virginia University, the University of Alabama at Huntsville, and the University of California at Los Angeles. Funding came from the National Science Foundation (NSF), the National Center for Atmospheric Research (NCAR), the National Oceanic and Atmospheric Administration/National Weather Service (NOAA/NWS), and the Atmospheric Environment Service, Canada (AES). The objective of VORTEX was to chase tornadoes in the southern Plains between April and June 15 of each year, in the hopes of intercepting about 30 supercell thunderstorms in the process of spawning tornadoes. VORTEX specifically sought to study factors in the environment that contribute to the spawning of a tornado (tornado genesis), tornado dynamics, and the distribution of debris from a formed tornado.

While 1994 produced few tornadoes, in 1995 the team collected data on 13 days, including June 2, when a tornado hit Dimmitt, Texas. Researchers used vehicles, aircraft, and weather balloons to effectively create a mobile mesonet to capture data from the event, including temperature, humidity, wind speed, and barometric pressure observations, as well as dramatic photography. The data and photography allowed researchers to attempt a numerical simulation of the event. VORTEX also represented the first use of a mobile Doppler radar, or "Doppler on Wheels," designed by Josh Wurman of the University of Oklahoma and constructed by the National Severe Storms Laboratory (NSSL) with support from the National Center for Atmospheric Research (NCAR).

VORTEX II was a second attempt of the same type of project, funded for $10 million in 2009/2010 by the National Science Foundation and NOAA, and equipped with 40 vehicles and 10 mobile radars. One notable accomplishment from the first year of VORTEX II was the intercept of a tornado in Wyoming on June 5, 2009. For the first time, the team was able to closely track the entire life cycle of a tornado. Researchers were even able to capture video of the inside of the tornado funnel, making it the most documented tornado in history.[1]

The data and observations generated by VORTEX and VORTEX II allow scientists to better understand how tornadoes form. Meteorologists may find such research projects of immense intrinsic value, and certainly tornadoes hold a certain fascination for many members of the public, including the authors of this book. However, while scientists may be content with academic knowledge, the average taxpayer is likely to look at a project like VORTEX II and want to know what the return on this investment will be. They want to see progress toward reducing the impacts of tornadoes on society, making it less likely that lives will be lost and communities devastated by tornadoes. In short, people want to know how these large research projects improve the well-being of individuals and communities. Societal impacts research attempts to answer this question, by examining how "academic" knowledge leads to practical knowledge that allows us to reduce the impact of tornadoes on society.

1.2.1. Doppler Radar

In the 1990s, the Department of Commerce embarked on a thoroughgoing modernization of the National Weather Service (NWS). The modernization included a professionalization of personnel (a shift to university-trained meteorologists), a reduction in the number of local weather forecast offices across the country, the implementation of a new Advanced Weather Information Processing System (AWIPS), and the construction of a nationwide system of new weather radars (WSR-88D) based on Doppler technology that

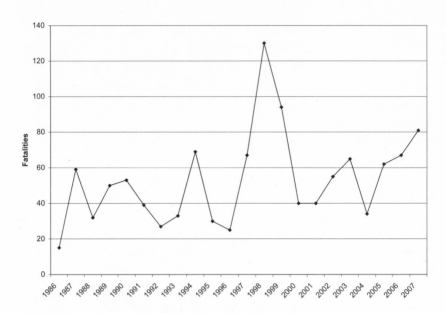

FIGURE 1.1. The modernization of the NWS and tornado fatalities

would be joined together for the first time in a true network, the NEXRAD network (Friday 1994). The NEXRAD system consists of 166 radars operated by the Departments of Commerce, Transportation, and Defense at a cost of $1.2 billion. The modernization was expected to yield a wide array of valuable new forecast products and weather services (Chapman 1992). One of the most visible expected benefits of the new Doppler radars was improved warnings, which would hopefully reduce the tornado death toll (Crum, Saffle, and Wilson 1998; Friday 1994).

Figure 1.1 displays the annual tornado fatality count for the years 1986–2007, and inspection of the time series suggests that the nation may not have purchased any reduction in tornado fatalities with the new Doppler radars. The WSR-88D radars were installed between 1992 and 1997, and if we consider fatalities in the six years immediately prior to and after the installation, we see that the two deadliest years in the sample were 1998 and 1999, immediately after the NEXRAD network was completed. Total fatalities *increased* by 70% from 248 in 1986–1991 to 424 in 1998–2003.

Did the new radars actually *cause* an increase in tornado fatalities, in contrast to the expected decline? Societal impacts research can answer this question. Doppler radar would only be expected to reduce the lethality of tornadoes when everything else is held constant: the famous *ceteris paribus* assumption from economics. A detailed analysis at the level of the individual tornado can attempt to hold these "other things" constant and estimate the effect of Doppler radar. In so doing, we can simulate an experiment with randomized trials with and without radar.

We have undertaken such an analysis (Simmons and Sutter 2005) and present updated results on this issue in Chapter 4. The analysis at the tornado level, controlling for a wide range of tornado and tornado-path characteristics, tells a very different story from the annual fatality totals in Figure 1.1.

1.2.2. Safe Rooms

For decades, residents of the Plains states have dug storm cellars for protection against tornadoes. Tornado protection has come a long way, however. In the 1980s and 1990s, engineers at Texas Tech University developed designs for "safe rooms" that can withstand the strongest tornado winds. Anyone who has visited the wind engineering lab at Tech and seen first-hand the contrast of a 2 × 4 fired from their wind cannon first at a traditional home wall and then at a safe-room wall has witnessed the amazing life-saving potential of these rooms. During the 1990s, safe-room principles were extended to a new generation of modern, underground tornado shelters. In principle, the safe-room design could be amplified to apply to entire homes. And yet as engineers were able to design shelters capable of protecting against tornadoes, homeowners in Tornado Alley stopped digging storm cellars. But shelters gained unprecedented attention when a family survived the May 3, 1999, tornado in Oklahoma City inside the only structure left standing from the destroyed home, their safe room. A total of 36 people perished and damages exceeded $1 billion, becoming the first tornado to have damages that high.

1.3. A Twister Gets Our Attention

The May 3 tornado attracted our professional interest as economists; we were both teaching at Oklahoma City–area universities that day. We were aware of and impressed with the new safe-room technology, but what was missing at the time was any hard analysis of the benefits versus costs of safe rooms or shelters. We knew that economics offered a set of research tools that would allow us to answer this valuation question. Our first collaborative research on tornadoes was to examine the benefits of shelters and safe rooms, and in the months after May 3, our efforts focused on figuring out how to assemble the data needed to perform a benefit-cost analysis, laying the groundwork for this now decade-long research project.

During the overnight hours of February 2, 2007, a fast-moving supercell tornado produced two F3 tornadoes in Lake County, Florida. The tornadoes struck The Villages retirement community and the rural Lake Mack area in the eastern part of the county. Lake County is also a popular destination for "snowbirds," who winter in Florida to escape the snow and cold of the north. News reports the next morning conveyed an all-too-familiar message: The tornadoes were killers, causing the deaths of 21 people, all in mobile homes.

Our research trip to Lake County a few weeks after the storm emphasized the vulnerability of these homes, as we saw relatively undamaged site-built homes adjacent to destroyed manufactured homes, and sites where the manufactured home had been blown away but the wooden stairs that led to the home were intact. We also met residents who were caught by surprise because they assumed tornado sirens would sound in advance of a tornado, but Lake County—like most Florida communities—does not have a siren system. The Lake County tornadoes embody what our research has identified as the four major vulnerabilities for casualties:

- Overnight (3 AM)
- Fall or winter months (February)
- Mobile homes (42% of the homes in Lake County)
- Southeast United States

Nonetheless, the event did have a silver lining: No fatalities occurred in any manufactured homes installed after the Department of Housing and Urban Development added the 1994 wind-load provisions to the HUD code for manufactured housing.

1.4. Mixing Meteorology and Economics

This study is an exercise in academic disciplinary trespassing. Our analysis is firmly grounded in the methods of economics, but the subject is outside of the traditional domain of topics studied by economists. Nobel Prize–winning economists Gary Becker and James Buchanan are famous for their interdisciplinary trespassing, using economic methods to study such topics as crime, the family, politics, and morality. This book fits into this broad tradition, and also can be thought of as an extension of environmental economics to weather and climate.

Interdisciplinary research must speak to (at least) two academic audiences. We hope we have written a book that will be of interest to atmospheric scientists and climatologists as well as to economists and social scientists interested in extreme weather. However, the message for these two audiences regarding tornado impacts will be different.

Meteorologists will be much more knowledgeable about tornadoes than the authors. Economists learn early about comparative advantage, and we will not pretend to be able to speak authoritatively on tornado genesis, multiple vortices, or the technical properties of weather radars. Any reader who doesn't already know about the physics of tornadoes isn't going to learn it here. But we believe that atmospheric scientists who study severe storms will find themselves drawn to the content of our impacts analysis and curious to learn about the determinants of tornado casualties, whether damage is increasing, and if tornado-warning false alarms create a "cry wolf" effect. In other words, meteorologists will be interested in this book as an example of societal impacts research. Many meteorologists and other physical scientists often consider societal impacts to be an "add on" to their research, but those who study cloud dynamics, for example, might be aware that in a world where funding for research is in shrinking supply, they must offer evidence of the value to society of their research. In this book, we will be studying the impacts side of the equation first and foremost. We will offer some innovative ways to attempt to estimate the value of meteorological research and the tornado-warning products of the National Weather Service. Through our analysis, meteorologists might discover some ways to think systematically about the value of their research to society, even if their areas of research are far afield from tornadoes.

Our book also attempts to reverse the standard direction of scientific research. Scientists focus on scientific advances first (which of course they see as tremendously valuable), with discussions of societal impacts a very distant second. We try to suggest some ways in which impacts can guide or direct research: We will start with information that we might need to reduce tornado impacts, and offer this as a target for research.

Economists would characterize our work as *applied*, meaning that we are using existing theoretical models and statistical techniques in our work. They will find this book interesting for very different reasons than meteorologists. Social scientists and the handful of economists who study natural hazards, and specifically extreme weather, will be interested in the substance of our study, the patterns we find, and how we address some recurring research challenges, as much of the data or many of the econometric problems that arise in studying tornado impacts also plague research on floods or hurricanes or forest fires. Others might simply find the book to be an interesting application of economics: Economists generally enjoy using their tools to study and understand the world, and applications of these tools to a new or different subject are often particularly interesting to them.

Finally, some economists might find our book interesting because it provides another piece of evidence on a few contentious and pervasive issues. For example, one area of current controversy is whether people adequately perceive and prepare for low-probability, high-consequence events. Natural disasters (and arguably financial and housing market disasters) are low-probability events, and many observers look at the impacts of the 2004 Indian Ocean tsunami and Hurricane Katrina in 2005 and conclude that as individuals and collectively, we fail to adequately prepare for disasters. Individual tornadoes do not have the regional or society-wide impacts of hurricanes, earthquakes, or other disasters, but they are clearly a high-consequence event for individual households. Thus, evidence on whether people seem to ignore tornado risk is relevant, but hardly decisive, in evaluating the prevalence of what is called low-probability event bias.

Another issue is the relationship between income inequality and risk. Economists generally find that safety is a luxury good, and that similar to other luxuries, safety is something we tend to consume much more of as we become wealthier. Recent research has looked at natural disasters to see if richer societies can afford more safety. The relationship between income and tornado impacts would be additional evidence on this question.

To conclude, we have tried to write a book that serves as a bridge between the pure science of tornadoes and the social implications these events have on the communities and people affected by them. No one discipline can adequately explain such a mysterious and often capricious phenomenon as a tornado; it is our hope that our research adds a significant element to this exciting field.

2

**TORNADO CLIMATOLOGY AND SOCIETY'S
TORNADO RISK**

2.1. Introduction

Tornado climatology refers to the *frequency* of tornadoes. Climatology is not social science, so it may seem odd for economists to begin a book on the economic and societal impacts of tornadoes with climatology. But many of the decisions people can make to reduce their vulnerability to tornadoes depend on an understanding of the likelihood of tornadoes, or climatology. Consider the following:

- Manufactured housing exhibits vulnerability to tornadoes, which the data will validate. Yet manufactured housing represents an affordable and increasingly comfortable housing option for many Americans (Beamish et al. 2001). How is a family concerned about tornado risk to decide whether to live in a mobile home? Are residents of tornado-prone states who nonetheless choose to live in a manufactured home simply ignoring or failing to perceive and appreciate weather risk, or balancing other important life goals against safety?
- Wind engineers have developed new shelters capable of protecting against even the most powerful tornadoes. The Federal Emergency Management Agency (FEMA) issued performance standards for new tornado shelters

in 1998 and included safe rooms in its (now-abandoned) National Mitigation Strategy. Tornado shelters and safe rooms are not cheap, and their benefits are tied to the risk of tornadoes. Are shelters worth the cost? An informed decision about shelter purchase requires data on tornado risk and how this risk varies across the nation.

■ The nation also invests in tornado research and technology, with the ultimate goal of helping the National Weather Service (NWS) forecast and warn for tornadoes. For example, in the 1990s, the United States invested $1.2 billion on a nationwide network of Doppler weather radars (the WSR-88D, or NEXRAD network). One of the expected benefits of Doppler radars was improved tornado warnings. In 2009, NOAA undertook a research study on tornadoes called VORTEX II (Verification of the Origins of Rotation in Tornadoes Experiment; see page 3). At a cost of $10 million, the project equipped 40 vehicles to intercept and observe the entire life cycle of a tornadic thunderstorm. Ultimately, the nation's return on these investments depends on the rate of tornado activity. The greater the threat, the greater the number of lives that can be saved by investments in research and technology to reduce the lethality of tornadoes.

Our examination of tornado climatology in this chapter focuses on the estimation of tornado frequency and differences in frequency across the United States. We do not plan to reinvent the wheel or refine previous spatial measures of tornado risk. Instead, we will consider the elements and limitations of these measures (and the tornado archives) that are relevant for the evaluation of tornado impacts. For example, measures of tornado frequency have been previously calculated (Schaefer et al. 2002, 1986) and used in analysis (Multihazard Mitigation Council 2005). Dividing the average annual damage area of tornadoes by land area of a region provides an estimate of the annual probability of tornado damage. Since not all tornadoes have equal destructive potential, F-scale adjusted frequency measures (F-scale is discussed in detail on the next page) have also been constructed, and limitations of the F-scale and existing tornado records have been discussed. This chapter will pull together prior observations by others and add several additional concerns arising from our analysis that affect the ability to assess the societal impact of tornadoes. We will address the following specific issues:

■ Like all analysis of extreme weather, tornado research depends on the quality of available records. It often employs a reasonably complete archive of U.S. tornadoes maintained by the Storm Prediction Center

(SPC), which includes records of tornadoes since 1950. The improved ability to document and record tornadoes and the challenge of climate change both raise the possibility that tornado frequency may be changing over time. If the underlying frequency of tornadoes is changing, the value of historical records for estimating risk today diminishes. Thus, we will see if we can identify a trend in tornado activity that is not an artifact of our changing ability to document tornadoes.

- Sixty years of records is a very short window of observation for events with a return period (for any given location) of thousands of years. Even if tornado frequency has not changed since 1950, the observed frequency of tornadoes based on available records may differ from the true frequency. In other words, we have no guarantee that the past 60 years reflect a "normal" rate of tornado activity. We must consider rates of tornado activity consistent with the observed record and consequently construct confidence intervals, as opposed to simply calculating 60-year averages. This chapter will also address the potential for a few more years of observation to affect our estimates of tornado frequency.

- Protective investments depend on the local tornado risk: For example, tornado risk is clearly greater in "Tornado Alley" than in New England. Although 60 years is a particularly short window in which to estimate differences in tornado risk across the United States, we will construct confidence intervals for state-specific measures of tornado frequency. The overall pattern of tornado incidence also differs across the nation, and these differences can substantially affect the risk to humans posed by tornadoes. In addition, the frequency of tornadoes, the pattern of activity throughout the day, the distribution of tornadoes by F-scale rating, and the concentration of risk across the year all affect the level of threat to people and property.

- The Fujita Scale (F-scale) was adopted by the NWS in the 1970s, and like any scale of damage for natural hazards, it has limitations. Although some of its weaknesses were addressed by the adoption of the Enhanced Fujita Scale (EF-scale) in 2005, some important limitations remain. First, the NWS did not begin using the F-scale until the 1970s, so ratings needed to be constructed retrospectively for earlier tornadoes. A potentially more serious limitation is that the F-scale is a damage measure, as opposed to an intensity measure. In sparsely populated areas, the potential to cause damage to sturdy buildings is limited. Strong and violent tornadoes (i.e., those rated F2 and stronger) in rural areas tend to be undercounted, and consequently, measures of tornado frequency based on archival records understate the threat to people and property.[1]

2.2. Tornado Incidence

The likelihood of a tornado occurring affects the value of investments made by households, businesses, and government to try to reduce the impact of tornadoes by protecting persons and property. Tornado frequency can be measured in one of two ways, the first based on the number of tornadoes occurring in a given area and timeframe, and the second based on the area of the tornado damage paths (not the amount of damage to property) during the time of the event. A damage area measure allows the estimation of a probability of tornado damage; if 10 square miles (mi^2) of tornado damage occurred over 50 years in 1,000 mi^2 of land area, then the probability of damage over the same period is approximately .01 = 10/1,000. Damage areas provide a better measure of incidence because tornado paths vary tremendously in length, width, and area. They can be smaller than a football field or large enough to cover 100 miles in length. The number of tornadoes depends on the area over which the count is made: The convention is to express the rate as tornadoes per 10,000 mi^2 land area per year. We will not report measures based on societal impacts here: for example, casualties or property damage per year. Impacts depend on the interaction of tornadoes and society, including the population of the area struck by tornadoes and the actions residents take to protect themselves. Impact-based measures do not convey a pure measure of risk due to nature alone, and may misrepresent the risk for persons who choose to live in the area. For example, imagine that a tornado slams through a desolate corner of West Texas every year. Nobody lives in the area, so the only impacts are damage to sagebrush and mesquite. We may accurately state that nobody ever died in a tornado in these parts. However, inferring from this observation that tornado risk is low could lead to the building of a mobile-home park on the site, with tragic consequences: The fatality count would mount quickly. Therefore, when we make decisions in locating vulnerable facilities or investing in protection, we should use risk measures based only on the frequency of tornadoes, not human interactions.

Tornadoes differ in intensity as well as in damage area. The F-scale measures tornado damage, not intensity, but damage correlates with intensity well enough that in the absence of any alternative the F-scale can be used as a measure of tornado strength. Tornado impacts derive mainly from stronger tornadoes. For example, the 977 tornadoes rated F0 or F1 in 2007 accounted for 4 deaths, while the 125 tornadoes rated F2 or stronger resulted in 77 fa-

TABLE 2.1. Summary Statistics of Annual Tornadoes

Variable	Total Tornadoes	Strong and Violent Tornadoes	Total Area (in Sq. Mi.)	Area of Strong and Violent Tornadoes (in Sq. Mi.)
Mean	874.0	188.3	334.3	273.3
Median	861.5	186	313.9	259.1
Standard Deviation	314.8	73.36	137.3	125.0
Minimum	202	80	91.06	58.33
Maximum	1823	389	768.3	644.5

talities, a difference of more than two orders of magnitude. A large number of weak tornadoes can inflate the likelihood of a tornado, but represents a relatively modest contribution to the true tornado threat. Thus, we tabulate a damage rate for all tornadoes, and then annual damage area for F2 and stronger tornadoes only.

We examine tornado climatology in order to assess the threat to life and property. Table 2.1 reports summary totals for the annual number of tornadoes in the United States over the period from 1950 to 2007. The source for these totals, as with most of the tornado figures reported in this analysis, is the tornado archive maintained by the SPC in Norman, Oklahoma.[2] The SPC archive reports one entry for each tornado that strikes a state, along with summary totals for tornadoes striking more than one state. We construct our analysis around the state tornado entry, because this represents the most disaggregated storm-level information available. Consequently, our analysis throughout the book reports state tornado segments, but we refer to them simply as tornadoes throughout the text; we use the term *state tornado* only when necessary to avoid confusion.[3]

Table 2.1 reports four measures of tornado frequency: the tornado rate per 10,000 mi^2 for all tornadoes and for strong and violent (F2 and stronger) tornadoes; and the damage area (in mi^2) per year for all tornadoes and for F2 and stronger tornadoes. A total of 50,691 tornadoes occurred in the contiguous United States since 1950,[4] with 10,291 rated F2 and stronger, or averages of 874 and 188 per year, respectively. The land area of the contiguous United States is just under 3 million mi^2, so the national annual tornado rate and the strong and violent tornado rate are 2.95 and 0.63 per 10,000 mi^2, respectively. The damage areas of all tornadoes and of F2 and stronger tornadoes since 1950 are 19,057 and 15,580 mi^2, respectively, with annual averages of 334 and

TABLE 2.2. Tornadoes by F-Scale Category, 1950–2007

F-Scale Category	Number of Tornadoes	Percentage of Tornadoes	Mean Length (in Miles)	Mean Width (in Yards)	Mean Area (in Mi²)
Unknown	1,843	3.63	1.36	8.98	0.06
0	21,417	42.24	1.03	9.84	0.03
1	16,522	32.59	3.00	22.57	0.17
2	8,119	16.01	6.54	43.95	0.67
3	2,199	4.34	13.70	94.39	2.62
4	540	1.07	23.41	150.5	6.88
5	64	0.13	30.27	219.8	9.86

273 mi², respectively. For the nation as a whole, tornado damage as a proportion of U.S. land area is .00011, and the annual proportion for F2 and stronger tornadoes is .000092. The return periods for tornado damage for the nation as a whole are 8,800 years for all tornadoes and 10,800 years for strong and violent tornadoes. The median of the annual totals is very near to the mean for each measure, indicating a symmetric distribution of annual totals. The reported totals exhibit considerable variation, with the number of tornadoes ranging from 202 to 1823, a difference of a factor of nine. The annual totals of strong and violent tornadoes range from 80 to 389. The damage areas of all tornadoes range from 91 to 768 mi², while the damage areas of strong and violent tornadoes range from 58 to 645 mi².

Table 2.2 reports the distribution of tornadoes by F-scale category. The F-scale was developed by Professor Theodore Fujita in 1972 and adopted by the NWS in 1973. The scale rates tornado damage on a six-point scale from 0 to 5, with 0 representing minimal damage and 5 "inconceivable" damage. Although Fujita proposed wind speed range estimates for the type of damage observed in a category, tornadoes are not rated based on measurement of their actual wind speed, but instead on an evaluation of the damage from the tornado. In 2005 the NWS switched to the EF-scale for rating tornado damage. The EF-scale maintains the 0 to 5 rating for tornadoes, and the numerical categories are intended to be consistent with the earlier F-scale ratings, so for simplicity we refer to all of the ratings as F-scale ratings. (We will return to the details of the F-scale and its enhancements later in this chapter.) Note that F-scale ratings were assigned retrospectively to tornadoes prior to 1975. Since the adoption of the F-scale, tornado ratings have been assigned by the NWS based on inspection of the damage paths; prior to 1975, ratings were based on available descriptions of the damage. Table 2.2 includes a category of unknown F-scale that accounts for 3.6% of tornadoes. In these cases (almost all from before the mid-1970s), the available description of damage was insufficient to assign the tornado an F-scale rating. However, given the lack of damage description, these tornadoes would appear to be weak.

Based on their F-scale ratings, tornadoes are characterized as "weak," "strong," or "violent," with F0 and F1 tornadoes in the weak category, F2 and F3 in the strong category, and F4 and F5 in the violent category. The majority of tornadoes are weak, with 42% rated F0 and 33% rated F1. The frequency of more powerful tornadoes drops rapidly, as 16% were rated F2, 4% rated F3, 1% rated F4, and 0.13% rated F5. The nation averages about one F5 tornado and ten F4 tornadoes per year. Table 2.2 reveals the variations in tornado damage paths across F-scale categories. The average damage path for F0 tornadoes was just over 1 mile in length, compared with an average path of 30 miles for F5 tornadoes. The average width of an F0 tornado damage path was 10 yards, compared with 220 yards, or 1/8 of a mile, for F5 tornadoes. This variation in both length and width of damage paths leads to an even more pronounced variation in area, from .03 mi^2 for F0 tornadoes to almost 7 mi^2 for F4 tornadoes and nearly 10 mi^2 for F5 tornadoes. Note that the mean characteristics of tornadoes with unknown F-scale values are very close to those of F0 tornadoes, suggesting that most of the tornadoes missing F-scale ratings were likely F0, with a few F1 tornadoes included in the SPC data as well.

The substantially greater damage areas of stronger tornadoes in Table 2.2 suggest that the distributions of tornadoes and damage areas by F-scale category differ substantially. Figure 2.1 displays both distributions. F0 tornadoes are most common, but F2 tornadoes cause the most damage, accounting for almost 1/3 of tornado damage. Less than 6% of tornadoes are rated F3 or stronger, and yet these tornadoes account for almost half of the damage area. It is important to remember that the NWS rates tornadoes based on the worst damage along the storm path and that tornadoes strengthen and weaken along their paths, so the proportion of area actually experiencing F4 or F5 damage will be less than the proportion reported in Figure 2.1. Comparison of the distributions of numbers of tornadoes and damage areas indicates the value of measures of frequency based on each. A state might experience a large number of weak tornadoes and appear to face great tornado risk based on the tornado rate, but have a much lower frequency based on damage areas. We will see that Florida fits this description.

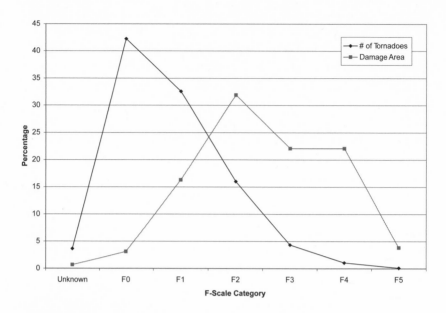

FIGURE 2.1. Distribution of tornadoes and damage area by F-scale

Concern over global warming has raised awareness of the potential for changing climatological normals. Tornadoes are one type of extreme weather that may increase with climate change, so we must be aware of and test for a change in the incidence of tornadoes over time, either nationally or regionally. Tornadoes are infrequent events; Schaefer et al. (2002) report that the highest estimated annual probability of tornado damage in the United States is 6×10^{-4} in central Oklahoma, which yields a return time of over 1,600 years. Estimation of tornado probabilities and particularly regional differences in these probabilities across the United States requires as many years of complete records as possible. At the same time, the changing climatology of tornadoes reduces the value of records from the past, since the frequency of tornadoes will begin to differ more and more. We consequently test for a change in the incidence of tornadoes since 1950.

Figure 2.2 displays the annual number of state tornadoes since 1950, and an increase over time is readily apparent. Fewer than 600 tornadoes were reported each year between 1950 and 1955, with a steady upward trend since then. New records for tornadoes were set in 1957, 1973, and 2004, with the record rising from 861 in 1957 to 1,104 in 1973 and 1,823 in 2004. Fitting a linear regression to the annual totals, as reported in Table 2.3, confirms the increase over time, with the total increasing by almost 16 state tornadoes per year, from an estimated 430 in 1950 to 1,300 in 2007, or a tripling of the annual total.

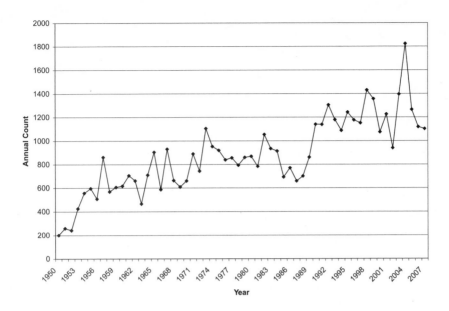

FIGURE 2.2. Tornadoes by year, 1950–2007

The total number of tornadoes is just one measure of tornado frequency, and we need to examine the other frequency measures before drawing any conclusions about climatology. It's possible that the total number of tornadoes appear to be increasing due to more effective reporting and documenting of tornadoes now than in the past. America has changed in many ways since the 1950s; today there are many more storm chasers, and video cameras and cell phones allow more effective documenting and reporting of tornadoes. The proportion of tornadoes reported and eventually entered into the SPC archive might be much greater now than in the 1950s; due to

TABLE 2.3. Time Trends in Annual Tornadoes, Damage Area

Measure of Tornado Activity	Time Trend	Constant
All Tornadoes	15.6 (11.39)	430 (9.51)
F2+ Tornadoes	−2.06 (4.04)	247 (14.64)
F4+ Tornadoes	−.159 (2.80)	14.9 (7.96)
Damage Area	−.963 (0.87)	361 (10.04)
F2+ Damage Area	−1.62 (1.63)	319 (9.90)

The table reports the results of a linear regression of each measure of annual tornado activity on a time trend term (equal to 0 in 1950) and a constant. Point estimates with absolute t-statistics based on robust standard errors in parentheses.

the increase in storm chasers and spotters, we may even be approaching 100% reporting, at least in tornado-prone states during the tornado season. In short, Figure 2.2 may simply reflect a change in the reporting, as opposed to the incidence, of tornadoes.

In reality, the improved efficiency of reporting has probably had the greatest impact on the reporting of short, weak tornadoes. A brief tornado in a rural area in the 1950s might have been seen only by a local farmer who was too busy working to report the tornado to the NWS; today, the same kind of storm could easily be broadcast live on local TV by storm chasers. On the other hand, longer-track, more powerful tornadoes causing damage to property would likely have been reported both in the 1950s and today. Figure 2.3 graphs the annual count of F2 and stronger tornadoes since 1950, and this count reveals no increase in tornado activity. The annual total of strong and violent tornadoes was between 100 and 150 in the early 1950s and in each year of the first decade of the twenty-first century. Interestingly, the total number of strong and violent tornadoes was greater between 1954 and 1976 than in the years since, with the annual total exceeding 200 eighteen times in the 23 years between 1954 and 1976, but only 4 times in the 31 years since. In fact, a time trend fitted to the count of F2 and stronger tornadoes reported in Table 2.3 indicates a statistically significant *negative* decline of about two strong and violent tornadoes per year, from around 250 in 1950 to around 130 in 2007; thus, the number of powerful tornadoes has declined by half, while total tornado reports have tripled. However, the high number of strong and violent tornadoes prior to 1976 may be an artifact of retrospective assignment of F-scale ratings to tornadoes by the NWS based on newspaper reports and other available evidence as opposed to damage surveys by NWS personnel. Researchers may have assigned higher retrospective ratings to tornadoes than they would have based on a contemporary damage-path survey. Still, even if we dismiss the decline in F2 and stronger tornadoes as a consequence of the retrospective assignment of early F-scale ratings, we still have no evidence of an *increase* in the frequency of strong and violent tornadoes over time.

Annual tornado damage provides additional evidence on trends in tornado frequency. Figure 2.4 graphs the annual damage area of all tornadoes and of tornadoes rated F2 and stronger. Neither series displays any evidence of an increase in tornado frequency; damage area in the 1950s averaged 325 mi^2 a year, with F2 and stronger tornadoes accounting for 280 mi^2, and the annual totals remain relatively steady over the period. Damage area fluctuates significantly from year to year, with extremes for all tornadoes of 768 mi^2

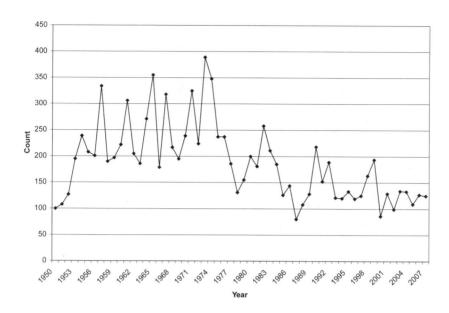

FIGURE 2.3. Strong and violent tornadoes by year, 1950–2007. Strong tornadoes are rated F2 or F3 on the F-scale of tornado damage, and violent tornadoes are rated F4 or F5.

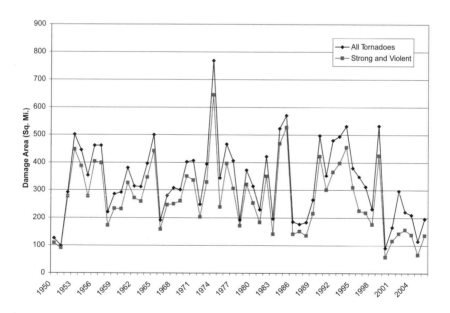

FIGURE 2.4. Tornado damage area by year, 1950–2007

TABLE 2.4. Distribution of Tornadoes by F-Scale, Pre- and Post-1976

F-Scale Category	Proportion, 1950–1975	Proportion 1976–2007
Unknown	.0806	.0141
0	.1973	.5354
1	.3615	.3080
2	.2701	.1049
3	.0698	.0301
4	.0181	.0069
5	.0027	.0006

Proportions of tornadoes occurring in each time period rated in each F-scale category

in 1974 and 91 mi² in 2000, and 645 mi² and 58 mi² for strong and violent tornadoes, respectively, in these same years. The correlation between the total damage area and the F2 and stronger damage area is +.98. Regression analysis in Table 2.3 confirms the lack of a trend in damage, with both total damage area and F2 and stronger damage area exhibiting a downward but statistically insignificant trend.

What does this analysis in total suggest about the trend in tornado frequency? While the annual number of tornadoes has significantly increased by about 900 state tornadoes per year since 1950, this is almost certainly a consequence of more complete reporting. Short-path, weak tornadoes that previously went undocumented are now reported and entered into official records with greater efficiency. If the increase in the total number of tornadoes were part of an increase in tornado activity, we should see increases in the other measures of frequency. But reported tornado damage area and F2+ damage area show no time trend, and a significant downward trend in the count of F2 and stronger tornadoes has been observed. This downward trend is likely due to the retrospective application of the F-scale to tornadoes from the 1950s and 1960s, compared with contemporary rating based on damage surveys since the mid-1970s.

Table 2.4 reports the distribution of tornadoes by F-scale category in the time periods of 1950–1975 and 1976–2007. Comparison of the distributions over these periods allows us to evaluate the overall impact of retrospective application of the F-scale ratings and changes in the reporting of tornadoes on apparent tornado risk. The proportions over the two periods differ considerably. Since 1976, F0 ratings have been most common, with over 53% of tornadoes receiving that rating. Prior to 1976, F0 was the third most common rating at 20% of tornadoes, trailing F1 at 36% and F2 at 27%. The proportion of tornadoes rated in each of the F3 through F5 categories before 1976 was more than double the proportion since 1976, so damage may have been over-rated retrospectively. When rating tornadoes retrospectively, experts have less information to guide them and may not be able to observe weaknesses in construction of buildings, for example, which may lead to a tornado being rated F3 instead of F2.

Because we believe that tornado frequency is unchanged since 1950, we can use the entire available archive to estimate tornado risk. Table 2.1 reported averages for annual measures of tornado activity. The United States has averaged 874 tornadoes per year, including 178 strong (F2 and F3) and 10 violent (F4 and F5) tornadoes. The annual probability of a tornado is the relevant measure of frequency we use to value investments to reduce societal vulnerability. The true probability, however, is not observed. The averages in Table 2.1 merely represent an estimate of the true rates based on 58 years of data. The "normal" number of tornadoes does not occur each year, and we observe active years for tornadoes (e.g., 2004) and years with less activity (e.g., 2002). With almost 60 years of data, we can hope that the above- and below-normal years will balance out, but we have no guarantee that the average of any of our measures of frequency over the period of 1950 to 2007 equals the true, unobserved frequency. It's possible that the last 60 years have been an unusually active (or inactive) period for tornadoes. The averages in the absence of a trend represent our best guess of the true tornado frequencies, but the potential exists for even a 60-year average to deviate from the true frequency, and we must account for the potential variation.

We can approach this problem by considering the 58 years of data as a sample drawn from the true distribution of tornado activity, and construct confidence intervals for the various measures of tornado rate. The exercise is simple but assumes significant practical importance. An investment that yields expected benefits in excess of costs based on the mean tornado rate may have greater costs than benefits for lower rates consistent with the 58-year averages, and an investment that is not worthwhile at the mean rate may be worthwhile at higher rates within the confidence interval. Confidence intervals for tornado activity are thus important in assessing the value to society of protective investments. A protective measure that yields benefits in excess of costs at the lowest tornado rates in the confidence interval is probably a wise investment.

Table 2.5 reports the 99% confidence intervals for our four measures of tornado activity using the mean and standard deviation of the annual totals for 1950 through 2007. The 99% confidence interval for the true, unobserved

TABLE 2.5. Confidence Intervals for Tornado Frequency, 1950–2007

Frequency Measure	C. I. Lower Bound	Mean	C.I. Upper Bound
Tornado Count	767.5	874.0	980.4
F2+ Tornado Count	163.5	188.3	213.1
Damage Area	287.5	334.3	381.2
F2+ Damage Area	230.7	273.3	305.8

TABLE 2.6. Confidence Intervals for Tornado Frequency, 1978–2007

Frequency Measure	C. I. Lower Bound	Mean	C.I. Upper Bound
Tornado Count	937.2	1068	1199
F2+ Tornado Count	109.2	146.1	183.0
Damage Area	243.4	313.6	383.9
F2+ Damage Area	183.7	247.7	311.7

number of tornadoes per year is 768 to 980, so the upper and lower bounds are within about 12% of the mean value. The confidence intervals for the other measures are slightly larger as a percentage of the mean. Tornado probability is inversely proportional to the damage area, so the ratio of damage areas indicates the ratio of the probabilities. Thus we see that the 99% confidence interval indicates that the overall tornado probability might differ by 14% from the mean level, and the probability of F2 and stronger damage might differ by almost 16% from the mean value. Overall, we can conclude that the true national tornado rates are likely to be within about 15% of the averages.

Although the lack of a trend suggests we can use all six decades of records to estimate confidence intervals, climatology normals are usually calculated using 30 years of data, so Table 2.6 reports confidence intervals constructed using annual totals from 1978 to 2007. As we reduce the number of years of observations used to construct the mean, the sample standard deviation tends to increase, and this will widen the confidence interval. The means of each measure of tornado frequency calculated over the last 30 years also differ from the 58-year means: Tornadoes per year are over 20% higher, F2+ tornadoes are 20% lower, and the mean damage areas are about 10% lower. We are particularly interested in the combined effect of these two factors on the upper bound of the confidence intervals. But as we see in comparing Tables 2.5 and 2.6, the upper bounds of the intervals for the damage area frequencies over the two periods are within 2% of each other. Therefore, using only recent data provides no indication that the maximum tornado probabilities consistent with observed means are significantly higher than for the entire 58-year sample. The lower bounds of the confidence intervals using 30 years of data are notably smaller than for the full period: 15% lower for the damage areas of all tornadoes and 20% lower for the damage areas of F2 and stronger tornadoes. Thus, an investment that is on the margin of cost effectiveness using tornado probabilities estimated over the entire period may not be worth undertaking if we consider only more recent data.

2.3. Tornado Risk Across States

Tornado risk is not equal across the nation: "Tornado Alley"[5] is much more at risk than New England or west of the Rocky Mountains. Due to the differences in tornado risk, some protective investments, regulatory responses, or actions by the NWS might be worthwhile only in high-risk parts of the country. To explore the variation in tornado risk, Table 2.7 reports our four measures of tornado frequency calculated for each of the contiguous 48 states. Since 1950, Texas has experienced the most tornadoes at 7,539, followed by Kansas and Oklahoma; all states have experienced tornadoes, but 8 have had fewer than 100 tornadoes, led by Rhode Island at 9, or about one every six years. Florida has the highest state tornado rate at 9.4 per 10,000 mi^2, followed by Oklahoma (at 7.7), Kansas (6.9), Iowa (6.3), and Illinois (6.0). The states with the lowest tornado rates are Arizona (.32), Washington (.24), Utah (.24), Oregon (.17), and Nevada (.11). The tornado rate in Florida is 85 times greater than the rate in Nevada, a difference of almost two orders of magnitude between the highest- and lowest-rate states.

Florida is not popularly viewed as the most tornado-prone state despite its top rank in tornado rate. The perception of Florida as a modest-risk state is borne out by the other measures of frequency. Florida ranks 29th in the probability of damage, 12th in the rate of F2 and stronger tornadoes, and 26th in the probability of F2 and stronger damage. Many Florida tornadoes are short and weak. The state with the highest annual probability of tornado damage is Mississippi at .000415; to place this number in perspective, the return period for tornado damage in Mississippi is about 2,400 years, so tornadoes are very infrequent events. In comparison, the annual probability of hurricane landfall in South Florida is greater than .1, or less than 10 years on average between hurricane landfalls, while the standard for flood risk is the 100-year flood plain, or a .01 annual probability event. Of course, the 2,400-year return period for Mississippi does not mean that a residence struck this year is automatically safe for the next two and a half millennia, because the risk is independent. The same town will sometimes be struck twice in a year, or even the same day.

The other states with the highest annual tornado probabilities are Arkansas, Oklahoma, Kansas, and Iowa. The states with the lowest annual probabilities are Nevada, Washington, Oregon, Arizona, and California. The annual probability of tornado damage in Nevada is 4.7×10^{-7}, which is a return period for damage of 2.1 million years. The annual probability of damage in Mississippi divided by the probability in Nevada is 887, so the difference between the highest and lowest tornado probability states is almost three orders of magnitude, compared with the two orders of magnitude difference for tornado rates.

Table 2.7 also reports the annual rate of F2 and stronger tornadoes per 10,000 mi^2 land area and the annual probability of damage from a tornado rated F2 or stronger. The states with the highest F2+ tornado rates are Oklahoma at 2.1, followed by Indiana, Mississippi, Arkansas, and Iowa, while Idaho, Arizona, Utah, Oregon, and Nevada have the lowest rates; indeed, Nevada has not had an F2 or stronger tornado since 1950. The states with the highest and lowest F2 or stronger tornado probabilities are Arkansas (.000360), Mississippi, Oklahoma, Iowa, and Kansas (highest), and Idaho, Oregon, Washington, Arizona, and Nevada (lowest). The annual probability of F2+ tornado damage in Arkansas is .000360, which is a return period of almost 2,800 years. Since Nevada has not experienced an F2 or stronger tornado since 1950, we have to take ratios of the most to second-least vulnerable states; these are 4,258 for the F2+ rate and 12,000 for the F2+ probability, so the range of risk for a strong or violent tornado exceeds the range of risk for all tornadoes.

The measures reported in Table 2.7 are averages over the entire state; however, vulnerability will vary across a state, with the variation potentially substantial in larger states. For instance, Texas ranks only 12th in tornado rate and 21st in tornado probability, yet Harris and Galveston counties are among the top 15 counties nationally in tornado rates (see Section 2.4 and Table 2.8). A more sophisticated method of estimating tornado probabilities breaks the contiguous United States into latitude and longitude grid boxes and assigns damage area to the grid boxes based on the reported latitude and longitude of each tornado. Such an approach has been applied by others (Schaefer et al. 1986, 2002; Multihazard Mitigation Council 2005), so we will not pursue it here. We are more interested in examining the societal impacts of tornado climatology than refining existing measures of frequency. Still, we can note

TABLE 2.7. Tornadoes by State

State	Tornadoes	Rate	F2+ Total	F2+ Rate	Probability	F2+Probability
Alabama	1,485	5.046 [10]	478	1.526 [6]	.000262 [7]	.000227 [7]
Arizona	210	0.319 [44]	13	0.020 [45]	.00000164 [45]	.00000030 [47]
Arkansas	1,427	4.725 [14]	533	1.765 [4]	.000410 [2]	.000360 [1]
California	362	0.400 [42]	23	0.025 [43]	.00000164 [44]	.00000100 [43]
Colorado	1,738	2.889 [27]	124	0.206 [35]	.0000180 [34]	.00000924 [35]
Connecticut	81	2.882 [28]	27	0.961 [19]	.0000614 [30]	.0000525 [25]
Delaware	57	5.029 [11]	13	1.147 [11]	.0000740 [27]	.0000152 [33]
Florida	2,940	9.400 [1]	347	1.109 [12]	.0000667 [29]	.0000432 [26]
Georgia	1,252	3.728 [17]	328	0.977 [17]	.000215 [12]	.000163 [12]
Idaho	183	0.381 [43]	10	0.021 [44]	.00000201 [43]	.00000098 [44]
Illinois	1,952	6.055 [5]	453	1.405 [8]	.000239 [8]	.000188 [11]
Indiana	1,167	5.610 [8]	377	1.812 [2]	.000299 [6]	.000272 [6]
Iowa	2,051	6.329 [4]	547	1.688 [5]	.000326 [5]	.000290 [4]
Kansas	3,285	6.923 [3]	595	1.254 [9]	.000329 [4]	.000276 [5]
Kentucky	655	2.843 [29]	231	1.003 [15]	.000120 [22]	.0000954 [22]
Louisiana	1,508	5.969 [6]	376	1.472 [7]	.000180 [13]	.000132 [14]
Maine	100	0.559 [40]	19	0.106 [38]	.00000407 [41]	.0000101 [42]
Maryland	274	4.833 [13]	39	0.688 [25]	.0000734 [28]	.0000246 [30]
Massachusetts	151	3.321 [23]	43	0.946[20]	.000110 [23]	.0000853 [23]
Michigan	909	2.759 [30]	261	0.792 [23]	.000163 [15]	.000109 [19]
Minnesota	1,400	3.032 [26]	247	0.535 [29]	.000129 [20]	.000101 [20]
Mississippi	1,596	5.866 [7]	490	1.801 [3]	.000415 [1]	.000355 [2]
Missouri	1,741	4.358 [16]	436	1.091 [13]	.000176 [14]	.000145 [13]
Montana	361	0.428 [41]	42	0.050 [41]	.00000639 [39]	.00000529 [38]
Nebraska	2,408	5.401 [9]	364	0.816 [22]	.000232 [11]	.000202 [8]
Nevada	75	0.118 [48]	0	0 [48]	.000000468 [48]	0 [48]
New Hampshire	81	1.557 [33]	22	0.423 [31]	.0000146 [36]	.00000683 [36]
New Jersey	138	3.208 [24]	28	0.651 [27]	.0000753 [26]	.0000233 [32]
New Mexico	508	0.722 [38]	38	0.054 [40]	.00000422 [40]	.00000112 [41]
New York	351	1.282 [35]	67	0.245 [33]	.0000405 [32]	.0000275 [28]
North Carolina	1,000	3.540 [21]	199	0.704 [24]	.000153 [17]	.000123 [16]
North Dakota	1,235	3.087 [25]	151	0.377 [32]	.0000332 [33]	.0000270 [29]
Ohio	859	3.698 [18]	242	1.042 [14]	.000161 [16]	.000131 [15]
Oklahoma	3,081	7.736 [2]	848	2.129 [1]	.000387 [3]	.000326 [3]
Oregon	95	0.171 [47]	3	0.005 [47]	.00000162 [46]	.00000094 [45]
Pennsylvania	658	2.531 [31]	177	0.681 [26]	.000132 [19]	.000112 [17]
Rhode Island	9	1.485 [34]	1	0.165 [37]	.0000987 [24]	.0000375 [27]
South Carolina	796	4.558 [15]	151	0.865 [21]	.000146 [18]	.000111 [18]
South Dakota	1,562	3.549 [20]	286	0.650 [28]	.0000867 [25]	.0000752 [24]
Tennessee	839	3.510 [22]	295	1.234 [10]	.000236 [9]	.000190 [10]
Texas	7,539	4.965 [12]	1,463	0.963 [18]	.000128 [21]	.0000988 [21]
Utah	112	0.235 [46]	9	0.019 [46]	.00000260 [42]	.00000201 [40]
Vermont	36	0.671 [39]	13	0.242 [34]	.0000137 [37]	.00000538 [37]
Virginia	534	2.325 [32]	106	0.462 [30]	.0000442 [31]	.0000244 [31]
Washington	94	0.244 [45]	15	0.039 [42]	.00000148 [47]	.00000067 [46]
West Virginia	116	0.831 [37]	25	0.179 [36]	.00000683 [38]	.00000376 [39]
Wisconsin	1,121	3.559 [19]	315	1.000 [16]	.000235 [10]	.000200 [9]
Wyoming	562	0.998 [36]	55	0.098 [39]	.0000156 [35]	.0000121 [34]

two limits of the grid box approach. First, the latitude and longitude coordinates of tornadoes in the SPC do not appear very accurate, a point we will return to in Chapter 3. If the coordinates of the tornadoes are of limited accuracy, we would probably be unable to improve on county-based measures of frequency. Second, the records at the county level are not only incomplete but biased in rural areas, a point we will return to later in Section 2.4.

We instead explore some time-series properties of our four measures. We considered in Section 2.2 the potential for a time trend in national tornado activity, and we will test for time trends in state tornado activity as well. We also consider the potential deviation of the observed tornado frequency in states from the true, unobserved rates. To do this, we construct annual totals of tornadoes, damage area, F2+ tornadoes, and F2+ damage area for states with at least 600 tornadoes over the period—just over 10 per year (26 states total). We begin by testing for time trends in the four measures of frequency in individual states. For the nation as a whole, we observe a statistically significant increase over time for all tornadoes and a significant decline over time for F2+ tornadoes, with no significant trends in the tornado probabilities. We find the same pattern for individual tornado-prone states. Twenty-three states exhibited statistically significant increases over time in total tornadoes; the only states without a significant positive time trend were Indiana, Michigan, and Oklahoma. Fourteen states featured statistically significant declines in F2 and stronger tornadoes.[6] A statistically significant decreasing trend in total or strong and violent damage area was observed only three times, with no significant trend in the other cases.

We turn next to confidence intervals for state tornado frequencies. Conceivably, two states could have similar annual tornado rates, but different year-to-year variance of activity. A larger variance for a given mean produces a larger confidence interval, and thus the true tornado rate may be greater in the high-variance state. To explore the potential for deviation across states, and particularly to identify states with large confidence intervals, we calculated a 99% confidence interval for each of the four measures for the 26 states with at least 600 tornadoes. Of course, since the mean levels differ across states, the upper bounds differ as well, and so a table of upper bounds of the confidence intervals would appear on the surface similar to Table 2.7. Instead of just reporting the upper bounds, Table 2.8 presents the ratio of the upper bound of the confidence interval to the mean. Two patterns can be observed. First, the confidence intervals are generally larger for the F2+ measures than for some measures for all tornadoes, which is not surprising since strong and violent tornadoes occur less frequently and thus exhibit more year-to-year

TABLE 2.8. State Tornado Frequency Confidence Intervals

State	Tornado Rate	F2+ Rate	Tornado Probability	F2+ Probability
Alabama	1.212	1.238	1.497	1.558
Arkansas	1.257	1.298	1.385	1.419
Colorado	1.242	1.333	1.626	1.761
Florida	1.177	1.301	1.578	1.724
Georgia	1.184	1.294	1.780	1.960
Illinois	1.264	1.283	1.355	1.411
Indiana	1.194	1.323	1.739	1.780
Iowa	1.198	1.265	1.397	1.424
Kansas	1.179	1.239	1.370	1.401
Kentucky	1.257	1.381	1.498	1.538
Louisiana	1.211	1.299	1.365	1.454
Michigan	1.181	1.319	1.758	1.757
Minnesota	1.216	1.306	1.449	1.553
Mississippi	1.209	1.249	1.523	1.569
Missouri	1.224	1.305	1.400	1.432
Nebraska	1.183	1.235	1.637	1.699
North Carolina	1.263	1.270	1.882	2.033
North Dakota	1.204	1.376	1.741	1.852
Ohio	1.242	1.287	1.523	1.578
Oklahoma	1.154	1.232	1.361	1.396
Pennsylvania	1.278	1.321	2.644	2.833
South Carolina	1.333	1.397	1.775	1.890
South Dakota	1.213	1.366	1.800	1.852
Tennessee	1.241	1.339	1.511	1.525
Texas	1.124	1.232	1.352	1.425
Wisconsin	1.190	1.260	1.542	1.597

The table reports the ratio of the upper bound of a 99% confidence interval for each measure of state tornado frequency, based on the variance of the annual totals, to the mean values reported in Table 2.6. Larger numbers indicate more uncertainty regarding the true annual rates consistent with the observed means. Confidence intervals were calculated for states with at least 600 tornadoes over the period 1950–2007.

variation. Second, the damage area measures exhibit larger confidence intervals than the comparable tornado rate measure. Damage area totals are substantially influenced by rare, long-track tornadoes.

The confidence intervals exhibit notable differences across states. The ranking of the ratio of the upper bounds of the confidence interval to the mean across states correlates fairly closely across our four measures of tornado frequency. We can identify some low- and high-variance states. Pennsylvania is a notable high-variance state, with an upper bound for damage area more than 2.5 times the mean, easily the largest divergence from the mean in any state. Pennsylvania also has a relatively large deviation of the upper bound from the mean for the tornado rate and the F2+ tornado rate. North Carolina, South Carolina, and Georgia are the other states with the most variance; the upper bound for the probability of F2+ damage is about double the mean in North Carolina and Georgia. The high year-to-year variation in tornado activity in these states may be due to hurricane-spawned tornadoes. The Dakotas also have relatively high variation with upper confidence interval bounds substantially greater than the mean. The states with the highest tornado totals, like Texas, Oklahoma, Kansas, and Iowa, have relatively small upper bounds, less than 20% larger than the mean for rates and about 40% larger than the mean for probabilities. Arkansas and Illinois have small upper bounds for tornado probabilities, but relatively large upper bounds for tornado rates. Substantial differences in annual variation, particularly for damage area, can be observed for neighboring states, as illustrated by the pairs of Kansas and Nebraska and Ohio and Pennsylvania.

How much might state tornado frequencies differ based on confidence intervals for the true rates? To evaluate this, we can compare rankings based on confidence interval upper bounds with the rankings based on means reported in Table 2.8. Consider South Carolina, which, based on its overall frequencies, is slightly above average in tornado risk, ranking between 15th and 21st in the four measures of frequency. However, South Carolina exhibits large year-to-year variation in activity, and thus ranks between 5th and 8th based on the upper bounds, or with risk very similar to Illinois. The most extreme case is Pennsylvania, which ranks 19th in tornado probability and 17th in the probability of F2 or stronger damage. The upper-bound of the confidence interval for damage in Pennsylvania is more than 2.5 times the mean, and thus Pennsylvania would rank 4th in upper-bound tornado probability. Therefore, although Pennsylvania and South Carolina are not high-risk states based on the average frequency, confidence intervals indicate that they may be high-risk states.

We next consider the distribution of tornadoes by month. Figure 2.5 displays the distribution of tornadoes and strong and violent tornadoes by month across the United States. More tornadoes occur in May (over 11,000) than any other month, accounting for 22% of all tornadoes through the year. June, April, and July are the next most active months, accounting for 20%, 13%, and 11% of tornadoes, respectively. Strong and violent tornadoes also peak in May at just over 2,300, or 21% of the total. F2+ tornadoes tend to occur somewhat earlier in the year, as 18% occur in April, 15% in June, 10% in March, and 7% in July. Also, compared to all tornadoes, larger percentages of strong and violent tornadoes occur in January, February, November, and December. Tornado season in the United States runs between April and July, when 65% of all tornadoes and 61% of strong and violent tornadoes occur.

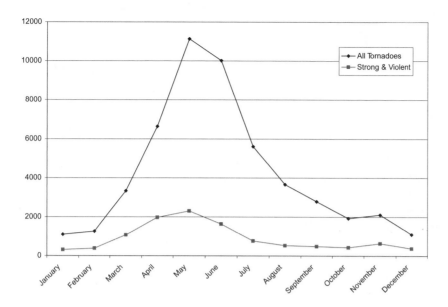

FIGURE 2.5. Tornadoes by month

We are also interested in the distribution of tornadoes across the year for individual states. We calculated tornado rates per 10,000 mi^2 for states per month, and multiplied the total number of state tornadoes in a month by 12 to maintain comparability with the annual rates. Table 2.9 displays the state with the highest tornado rate for each month. We clearly see that peak

TABLE 2.9. States with Highest Tornado Rate by Month

Month	State	Rate (per 10,000 mi²)
January	Florida	5.76
February	Mississippi	7.23
March	Florida	9.86
April	Oklahoma	18.35
May	Oklahoma	34.98
June	Nebraska	21.94
July	Delaware	19.06
August	Rhode Island	11.88
September	Florida	11.74
October	Florida	7.94
November	Alabama	9.38
December	Mississippi	5.12

tornado season actually occurs at different times of the year in different regions. Florida has the highest tornado rate overall and the highest rate in four different months: January, March, September, and October. Table 2.10 also displays the annualized rate in the peak month for each state. The peak-rate states during the winter and fall are generally in the Southeast; in addition to Florida, Mississippi and Alabama have the highest rates in February, November, and December. States in Tornado Alley have the highest rates during the prime tornado season. Oklahoma has the highest tornado rate in April and May, and Nebraska has the highest rate in June. Finally, the highest rates in July and August occur in Delaware and Rhode Island, two states that seem unlikely candidates to lead the nation in tornado rate in any month; indeed, no tornadoes occurred in Rhode Island in any month prior to August. The peak rates in the prime season months are significantly higher than the peak rates in the fall or winter months. For instance, the tornado rate in Oklahoma is almost 35 in May and over 18 in April, and Nebraska's rate in June is almost 22; the peak rates for October through March are less than 10 per 10,000 mi^2.

Table 2.10 lists each highest and lowest monthly tornado rate for the year by state, along with the month(s) in which the rates occur. The modal peak month for states is June, with 17 state peaks, including most of the northern Plains and Great Lakes states. Peak rates for 11 states occur in July, generally Northeastern states in addition to Pennsylvania, Maryland, and Delaware. Seven states have peaks in April and May, with the April peak states generally in the Southeast and the May peak states in the southern Plains states (Texas, Oklahoma, Kansas, and Missouri, plus Illinois). Three states have peaks in August, two in September (South Carolina and Virginia, which may be due to hurricane-spawned tornadoes), and one state (California) in March. Twenty-four states have at least one month with no recorded tornadoes since 1950; Rhode Island has experienced tornadoes in only three months, August, September, and October, even though the state has the highest August rate of any state. Nevada ranks last in tornado frequency by each of the four measures in Table 2.7, and yet has had tornadoes in every month of the year. Many states have more than one month with no tornadoes, which complicates description of the minimum (trough) months. Among states with a single lowest-rate month, the trough most often occurs in winter, with 20 states having a trough in December, January, or February, compared to 10 states with troughs in other months. The peak rate of activity exceeds Florida's tornado rate for the entire year in 25 states, and 10 states have a peak tornado rate exceeding 15 tornadoes per 10,000 mi^2. So

TABLE 2.10. Variation in Tornado Rates by Month by State

State	Max Tornado Rate	Month	Minimum Tornado Rate	Month(s)	Difference Rates	Max to Average
Alabama	10.07	April	2.000	August	8.07	2.00
Arizona	0.819	August	0.055	December	0.76	2.57
Arkansas	11.88	April	0.676	August	11.21	2.51
California	0.862	March	0.146	June	0.72	2.15
Colorado	12.67	June	0	1,11,12	12.67	4.38
Connecticut	8.968	July	0	1,2,3,11,12	8.97	3.11
Delaware	19.06	July	0	2,12	19.06	3.79
Florida	17.00	June	4.489	November	12.51	1.81
Georgia	7.97	April	1.215	October	6.75	2.14
Idaho	1.175	June	0	1,11	1.18	3.08
Illinois	17.35	May	0.707	February	16.64	2.86
Indiana	15.34	June	0.692	January	14.65	2.74
Iowa	21.15	June	0.074	December	21.07	3.34
Kansas	28.60	May	0.076	January	28.53	4.13
Kentucky	7.968	April	0.521	December	7.45	2.80
Louisiana	11.54	April	2.660	July	8.88	1.93
Maine	2.279	July	0	1,2,3,12	2.28	4.08
Maryland	14.61	July	0	December	14.61	3.02
Massachusetts	14.51	July	0	1,2,4	14.51	4.37
Michigan	7.321	June	0	December	7.32	2.65
Minnesota	12.48	June	0	1,2,12	12.48	4.11
Mississippi	11.25	April	1.588	July	9.66	1.92
Missouri	12.94	May	1.111	January	11.83	2.97
Montana	1.848	July	0	1,2,11,12	1.85	4.32
Nebraska	21.94	June	0	February	21.94	4.06
Nevada	0.339	June	0	December	0.34	2.88
New Hampshire	7.613	July	0	1,2,3,4,11,12	7.61	4.89
New Jersey	11.16	July	0	1,12	11.16	3.48
New Mexico	2.608	June	0	1,2,11	2.61	3.61
New York	3.812	July	0	1,12	3.81	2.97
North Carolina	7.730	May	0.467	December	7.26	2.18
North Dakota	13.71	June	0	1,2,12	13.71	4.44
Ohio	10.23	June	0.155	December	10.07	2.77
Oklahoma	34.98	May	0.362	January	34.62	4.52
Oregon	0.323	June	0.021	1,2	0.30	1.89
Pennsylvania	7.063	July	0.046	February	7.02	2.79
Rhode Island	11.88	August	0	1,2,3,4,5,6,7,11,12	11.88	8.00
South Carolina	8.314	September	1.237	October	7.08	1.82
South Dakota	17.26	June	0	1,2,11,12	17.26	4.86
Tennessee	9.889	April	0.402	September	9.49	2.82
Texas	17.93	May	1.099	January	16.83	3.61
Utah	0.604	August	0	October	0.60	2.57
Vermont	2.237	July	0	1,2,4,11,12	2.24	3.33
Virginia	5.069	September	0.105	December	4.96	2.18
Washington	0.746	May	0	February	0.75	3.06
West Virginia	2.664	June	0.086	February	2.58	3.21
Wisconsin	12.46	June	0	February	12.46	3.50
Wyoming	5.007	June	0	1,2,3,10,11,12	5.01	5.02

even though Florida has the highest risk of tornadoes year-round, we see that tornado risk is much greater, at least during a part of the year, in many other states, and that all of those states with the exception of Delaware have rates exceeding the annual mean. The most tornado-prone states tend to have pronounced peaks to their tornado season, as the correlation between a state's overall tornado rate and its peak monthly rate is +.85.

The final two columns in Table 2.10 report measures of the variation in tornado rate across the year for each state. The variation in activity across the year, or the extent of a well-defined tornado season, can potentially affect the threat to people and property posed by tornadoes. Residents of a state with a very pronounced tornado season can focus their awareness in these months and thus be prepared when tornadoes strike. Consider in the limiting case a state with high risk but where all tornadoes occur between April and July. Residents must be alert for tornadoes during the season, but can relax (or focus their attention on life's other risks) during the other months. By contrast, if the same number of tornadoes occurs evenly throughout the year, residents would need to be alert for all 12 months to avoid being caught off guard. Table 2.10 reports the difference between the maximum and minimum monthly rates for each state and the maximum monthly rate divided by the annual rate from Table 2.7. The differences between the maximum and minimum monthly rates track the maximum rates closely, since the minimum monthly rate is either zero or very close to zero. The differences between the maximum and minimum monthly rates and the annual tornado rates are highly correlated (+.80). The ratio of the maximum rate to the annual rate has a median of just over 3 and exhibits notable deviation across states. Outlier values of this ratio are observed in states with low rates, while the ratio ranges from just under 2 to about 4.5 in states with above-average tornado rates. Among high-risk states, the highest ratios are observed in the Tornado Alley states of Oklahoma, Kansas, and Nebraska, at over 4.0. Southeastern states like Florida, Alabama, Mississippi, and South Carolina have ratios less than 2. Again we see that Tornado Alley has a very pronounced season with rates well in excess of the average for the year, while the tornado threat is more consistent throughout the year in the Southeast. As another way of measuring a state's tornado season, we determined the four consecutive months with the highest tornado counts and tallied the percentage of state tornadoes occurring during this "season." In the United States, the season is March through June and accounts for just over 65% of all tornadoes. The pronounced season in Tornado Alley accounts for four out of five tornadoes in Oklahoma (81%), Kansas (80%), Nebraska (86%), South

Dakota (93%), North Dakota (97%), and Minnesota (87%). By contrast, the season accounts for less than half of tornadoes in Alabama (49.6%), Mississippi (49.8%), Florida (47%), Louisiana (47%), and South Carolina (48%). Tornado awareness in the Plains can be focused during the season, but must effectively be year-round in the Southeast.

2.4. Population Density Bias and County Tornadoes

Several years ago in the course of our research, we calculated tornado rates for counties in Oklahoma and noticed that the three most populated counties in the state, Oklahoma, Tulsa, and Cleveland, had three of the highest tornado rates. We later observed a similar pattern in other states. Such a coincidence would be an unfortunate societal vulnerability, as a disproportionate share of tornadoes would occur where the most people live, endangering more people. If most tornadoes instead struck sparsely populated ranch country far from any major towns, fewer people would be killed or injured. Rather than being an unfortunate vulnerability, however, the high tornado counts are more plausibly a consequence of the population. The probability that a tornado will be observed, reported, confirmed, and entered into the tornado archive is likely higher in more densely populated counties. Such a population density bias has been observed by other researchers (Ray et al. 2003; Anderson et al. 2005); the exact magnitude of the bias remains uncertain, yet the potential undercounting of tornadoes in sparsely populated areas affects any estimation of tornado frequency.

To explore the potential population density bias in the tornado records, we created county tornado counts and calculated the annual tornado rate per 10,000 mi^2. The tabulations allow us to consider the counties that have been struck by the most tornadoes since 1950. Table 2.11 reports the top 30 counties (approximately 1% of U.S. counties) in total reported tornadoes and the annual tornado rate per 10,000 mi^2. Weld County, Colorado, has experienced the most tornadoes since 1950 at 238, or about four per year. Harris County, Texas (Houston) ranks second at 208, or about one tornado a year more than the county that ranks third, Polk County, Florida, at 152. Another 9 counties (for 12 total) have been struck by 100 or more tornadoes. The top two counties based on the tornado rate stand out as outliers—Pinellas, Florida (St. Petersburg) and Galveston, Texas, at 49.3 and 45.9, respectively—have about double the rate of the third-ranking county, Oklahoma County, Oklahoma, at 23.9. Pinellas and Galveston are densely populated coastal

TABLE 2.11. The Most Tornado-Prone U. S. Counties, 1950–2007

	County	Tornadoes		County	Tornado Rate
1	Weld, Colorado	238	1	Pinellas, Florida	49.3
2	Harris, Texas	208	2	Galveston, Texas	45.9
3	Polk, Florida	152	3	Oklahoma, Oklahoma	23.9
4	Adams, Colorado	147	4	Lee, Florida	22.9
5	Palm Beach, Florida	144	5	Sarasota, Florida	22.3
6	Hillsborough, Florida	125	6	Hall, Nebraska	22.1
7	Hale, Texas	124	7	Union, New Jersey	21.8
8	Washington, Colorado	119	8	Hale, Texas	21.3
9	Pinellas, Florida	115	9	Adams, Colorado	21.3
10	Broward, Florida	110	10	Escambia, Florida	21.1
11	Dade, Florida	109	11	Lafayette, Louisiana	21.1
12	Lee, Florida	106	12	Orleans, Louisiana	21.0
13	Jefferson, Texas	100	13	Harris, Texas	20.7
14	Oklahoma, Oklahoma	98	14	Hillsborough, Florida	20.5
15	McLean, Illinois	97	15	Nueces, Texas	19.4
16	Volusia, Florida	95	16	Jefferson, Texas	19.1
17	Caddo, Oklahoma	95	17	Scott, Iowa	18.4
18	Kay, Oklahoma	95	18	Carteret, North Carolina	18.2
19	Nueces, Texas	94	19	Johnson, Texas	18.2
20	Lea, New Mexico	91	20	Milwaukee, Wisconsin	17.8
21	Okaloosa, Florida	90	21	Kay, Oklahoma	17.8
22	Lamb, Texas	90	22	Harrison, Mississippi	17.2
23	Sherman, Kansas	89	23	Arapahoe, Colorado	17.2
24	Kit Carson, Colorado	88	24	East Carroll, Louisiana	16.8
25	Laramie, Wyoming	88	25	Tarrant, Texas	16.8
26	Baldwin, Alabama	85	26	Okaloosa, Florida	16.6
T27	Castro, Texas	84	27	Scott, Missouri	16.4
T27	Tarrant, Texas	84	28	Gregg, Texas	16.4
T29	Pasco, Florida	82	29	Castro, Texas	16.1
T29	Lincoln, Nebraska	82	30	Marion, Indiana	16.1
T29	Carson, Texas	82			

counties in hurricane- and tropical-storm-prone states, which may explain their ranking. In total, 16 counties have a tornado rate of more than double the highest state rate (Florida at 9.4) from Table 2.7; it's not surprising that some locales have a much greater rate of activity than entire states. We also see the importance of differences in vulnerability within a state. Two Colorado and nine Texas counties rank among the top 30 in tornado rates even though these states rank 27th and 12th nationally in tornado rate, respectively, and Union County, New Jersey, ranks 7th even though New Jersey ranks 24th. Some parts of states with average tornado risk overall may face a very elevated rate of tornadoes.[7]

We then regress the county tornado rate on the average county population density from the six censuses from 1950 to 2000. To provide a thorough test, we wish to include a large number of states to ensure that we can distinguish between population density and differences in tornado frequency that happen to correlate with population. For instance, the western parts of Texas, Oklahoma, Kansas, and Nebraska are sparsely populated and have lower tornado rates than the eastern parts of these states. The difference may in fact be due to a difference in tornado frequency, and yet simply appears to be the result of population density bias. If we observe population density bias across a large number of states, it will be stronger evidence that the tornado rates truly vary due to population density. However, when we include different states in one sample, we are combining states with different underlying tornado frequencies. To control for the different rates across states, we include state dummy variables in the regression analysis. These state variables control for the fact that Illinois and Oklahoma do not have the same tornado rates, for example. If more densely populated counties in each state have higher tornado rates than less populated counties, the population density variables will capture this, and the state variables will control for differences in rates across states. State dummy variables control imperfectly for differing tornado rates, because it would be pure coincidence for differences in tornado climatology to coincide with boundaries of political geographies. State dummy variables provide a reasonable approach for our analysis here.

Table 2.12 reports the regression analysis of the annual tornado rate per 10,000 mi^2 measure of tornado frequency. The models include state dummy variables, but they are not reported in Table 2.12 because their coefficient estimates are immaterial for testing population density bias. The regressions use all counties in 12 tornado-prone states, chosen to ensure that (essentially) all counties in the data set have experienced tornadoes. The table reports three regression specifications. The first uses only a constant and the county population density (in thousands of persons per mi^2). We find support for a population density bias, as population density attains significance with a positive sign. Population density varies greatly across the United States, from less than one person per mi^2 to over 10,000 persons per mi^2, and thus the coefficient on density is small. Also, interpretation of the coefficient in this first specification is problematic, since we will see that the relationship is not linear. The second column includes density and density squared to control for a nonlinear relationship between density and tornadoes. We

TABLE 2.12. Population Density Bias and Tornado Rate

Intercept			2.75**
			(6.55)
Density	.220**	.00119**	
	(7.53)	(11.87)	
Density^2		−1.89E-8**	
		(10.65)	
Density10–50			2.60**
			(10.28)
Density50–100			4.53**
			(12.71)
Density100–500			6.62**
			(16.32)
Density500–1,000			9.34**
			(9.00)
Density1,000+			9.97**
			(10.54)

The dependent variable is the annual rate per 10,000 sq. mi. Population density is in persons per square mile. The models include all counties in states east of the Rocky Mountains, and include state dummy variables, which are not reported here and a constant term.
** Indicates significance from zero at the .01 level in a two-tailed test.

see that population density significantly increases tornado rates, while density squared significantly reduces the rate. The marginal effect of density is positive throughout the observed range of county populations, so the results show that the reported tornado rate increases with population density, but at a decreasing rate. The third column includes dummy variables for population density intervals, specifically for densities of 10 to 50 persons/mi^2, 50 to 100 persons/mi^2, 100 and 500 persons/mi^2, 500 and 1,000 persons/mi^2, and more than 1,000 persons/mi^2, with an average population density of fewer than 10 persons/mi^2 as the omitted category. The coefficient for the density interval variables indicates the increase of the tornado rate relative to a county in the category of fewer than 10 persons/mi^2. This specification provides an intuitive measure of the potential magnitude of the substantial population density bias observed in the records. The tornado rate in the most populous counties, with a population density in excess of 1,000 persons/mi^2, is almost 10 tornadoes per year higher than in sparsely populated counties, which exceeds the highest observed statewide tornado rate. The estimated tornado rate increases with each interval, and the differences from one interval to the next (except for the 500–1,000 and 1,000+ categories) are statistically significant.

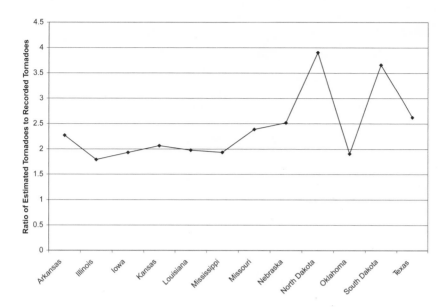

FIGURE 2.6. Ratio of estimated to reported tornadoes in 12 tornado-prone states

How much of a difference in a statewide tornado rate results from the population density bias? To investigate this, we apply the population density intervals specification and the state effects for each of the 12 tornado-prone states. We estimate the number of tornadoes that would have been recorded in the state if all counties had a population density in the maximum interval.[8] We then compare the estimated totals to the actual totals to measure the potential magnitude of the population density bias. Figure 2.6 reports the ratios for the 12 states. For 6 of the 12 states, the ratio is very close to 2, so we estimate that about half of all tornadoes have been recorded over the period. The smallest ratio is 1.9 in Illinois, where almost 53% of tornadoes were estimated to have been reported. The largest ratios of 3.9 and 3.7 are in North Dakota and South Dakota, with almost 3 out of 4 tornadoes possibly going unreported due to population density bias. Texas has the third-highest ratio at 2.7, so our estimate is that only about 37% of tornadoes in Texas have been reported. Nebraska, Missouri, and Arkansas have ratios of between 2.3 and 2.5.

The difference in the ratios of expected to actual tornadoes across states seems to be related to population patterns across the states. The proportion of a state that is sparsely populated is more relevant in explaining unreported tornadoes than state population density overall. Specifically, for each of the 12 states we calculate the percentage of state land area in counties with an average population density of fewer than 10 persons/mi^2. Over 70% of the area of North and South Dakota is sparsely populated by this measure, and these are the states where almost 3 out of 4 tornadoes may not have been recorded. By contrast, every county in Illinois and Iowa has a population density in excess of 10 persons/mi^2, and less than 5% of the area in Arkansas, Louisiana, Mississippi, and Missouri is so sparsely populated. Nebraska and Texas have 57% and 43% of their land area sparsely populated, respectively, and have ratios estimated to actual tornadoes of just over 2.5. Sparse population does not explain all of the variation: For example, 62% of Kansas is sparsely populated and yet only about half of the tornadoes in the state appear to have gone unreported, and Arkansas and Missouri are less than 5% sparsely populated, but have ratios in excess of 2.0. Local factors like the willingness of the population to report tornadoes or possible differences in the number of storm spotters could be relevant, or alternatively the efficiency of the tornado-reporting process could differ across states.

The tornado rate is just one measure of tornado frequency, and we saw in Section 2.2 that the time trend of the count of tornadoes provides a misleading signal of increasing tornado frequency. The question then immediately arises as to whether the population density bias in the tornado count is also due to more complete reporting of weak, short-path tornadoes in more populated areas. To investigate this, we tabulate total county tornado damage over the period 1950–2006. A proviso is necessary on the construction of this measure. The SPC archive reports damage area for state segments, so we have only the total damage-path area for a multicounty tornado; in this case, we assign an equal proportion of the damage area to each county. Obviously this procedure will misassign damage area for individual tornadoes, because the actual tornado damage path will not be evenly divided among counties. However, more than 4,000 tornadoes in our data set struck more than one county, so some balancing-out of the errors in assigning damage area can be expected. Also, there seems to be no reason to expect that the resulting error in apportioning damage area should be biased toward more or less sparsely populated counties. As long as errors in damage-path assignment are uncorrelated with county population density, we should be able to validly test for a population density bias in damage area.

We estimate the same regression models for county damage area as for the county tornado rate (including state dummy variables), and Table 2.13 presents the results. We observe population density bias in damage area as well as in tornado count. In the first specification, population density significantly increases damage in the total damage area.[9] The second column

TABLE 2.13. Population Density Bias and Tornado Damage Area

Density	4.07E-4*	7.20E-4*	
	(2.46)	(1.76)	
Density^2		−4.49E-6	
		(0.83)	
Density10–49.9			.00711**
			(4.12)
Density50–99.9			.00738**
			(3.05)
Density 100–499.9			.0116**
			(4.23)
Density500–999.9			.182**
			(2.60)
Density1,000+			.0168**
			(2.65)

* and ** indicate significance from zero at the .10 and .01 levels in a two-tailed test, respectively.

includes population density squared to test for a nonlinear relationship, and the point estimate for density is increased while population density squared has a negative point estimate that fails to attain significance. We see some evidence of a declining marginal effect of density on damage area, but that the quadratic relationship is not as strong as in the tornado rate. The population density interval variables all increase the damage area and attain statistical significance. The differences between the intervals are not all significant. Damage area is approximately equal for counties with average densities between 10 and 50 and between 50 and 100 persons/mi^2, and for counties with densities between 500 and 1,000 and more than 1,000 persons/mi^2. Total damage area increases with density, indicating that the bias is not merely for weak tornadoes but rather that more substantial tornadoes are not being reported in sparsely populated counties, and measures of tornado frequency fail to accurately measure risk in sparsely populated counties and states.

We next examine whether the population density has changed over time. Given the increase in the number of reported tornadoes, we would expect that the recent tornado rate will be higher than the earlier rate at all population densities. But the societal changes that have improved the efficiency of reporting tornadoes over time should probably have reduced the population density bias. Even in the 1950s, tornadoes in densely populated areas were probably more likely to be directly observed by residents or to cause damage to buildings and thus get reported, while tornadoes in rural areas were

TABLE 2.14. Is the Population Density Bias Diminishing?

Average Population Density Interval	Tornado Rate 1950–1979	Tornado Rate 1980–2007	Percentage Change in Tornado Rate
0–9.99	1.94	3.34	+71.56
10–49.99	3.58	5.53	+54.43
50–99.99	4.06	6.08	+49.79
100–499.99	5.03	7.00	+39.12
500–999.99	6.70	7.79	+16.22
1,000+	6.92	11.11	+60.58

substantially less likely to be recorded. Today, the increased number of storm spotters and storm chasers, Doppler radar, video cameras, cell phones, and survey teams should have a relatively greater impact on reporting tornadoes in less populated areas. If so, population density bias should be smaller over the more recent period than in the earlier decades.

To investigate this, we calculate county tornado rates over the first and second parts of our sample, or 1950–1979 and 1980–2007, respectively. Table 2.14 reports the tornado rates for counties averaged over population density intervals for 1950–1979 and 1980–2007. In each density interval, the tornado rate in the more recent period exceeds the rate in the earlier decades, but the percentage increase in the rate between two periods is smaller in the higher population density intervals, except for counties with population densities in excess of 1,000 persons/mi^2. The tornado rate increased 72% for the least populated counties, 54% for densities between 10 and 50 persons/mi^2, 50% for densities between 50 and 100 persons/mi^2, 39% for counties with densities between 100 and 500 persons/mi^2, and 16% for counties with densities between 500 and 1,000 persons/mi^2. The tornado rate for counties with a population density greater than 1,000 persons/mi^2 increased by 61%, exceeding the increase for all but the least populated counties. While this large increase for the most densely populated counties appears to contradict the hypothesized reduction in population density bias over time, the large percentage increase is actually an artifact of the data. Counties with population density greater than 1,000 are clearly urbanized, and include many independent cities and small counties. Fifty-six counties included in Table 2.14 have a land area of less than 100 mi^2, and 45 of these counties have a population density greater than 1,000 persons/mi^2. In addition, 43% of the most densely populated counties have areas of less than 100 mi^2. A small land area results in very large percentage increases (decreases) when tornadoes occur (do not occur), inflating the percentage increases in this category. To control for this, we recalculated the average tornado rates for counties with a density greater than 1,000 and land area of at least 100 mi². When we do this, we observe a 26% increase in the average tornado rate, which although greater than the increase for counties in the 500–1,000 persons/mi² range, is smaller than the increase observed in all of the lower density intervals.

2.5. Are Strong Tornadoes Underrated in Sparsely Populated Areas?

In 2005, the NWS introduced the Enhanced Fujita (EF) Scale in response to concerns about the limitations of the Fujita Scale (see the Texas Tech Fujita Scale Report, 2006). The enhancements of the Fujita scale essentially involve new criteria for use in the surveys that rate tornado damage. The six-point (0 to 5) scale is maintained for continuity with previous F-scale ratings. The Enhanced Fujita Scale includes a 28-point scale for damage assessment, and these new, detailed criteria will hopefully increase the consistency and ac-curacy of the 0 to 5 ratings. It was previously possible for different survey teams to essentially assign different ratings to the same damage; although subjectivity in damage assessment still exists, its influence has hopefully been reduced through the use of the EF-scale.

The EF-scale, however, does not address what is probably the major limi-tation of the F-scale in evaluating tornado risk. The F-scale was, and the EF-scale continues to be, a damage scale and not technically an intensity scale. (This is not a new observation; see Doswell and Burgess 1988.) F-scale ratings, despite this limitation, serve well in allowing researchers to control for the magnitude of the event. Research on tornado impacts is advantaged relative to research on, say, flash floods, which lacks a good measure of event intensity. We will follow the lead of other researchers who use the F-scale as a substitute for the intensity or strength of a tornado in the absence of a direct, better measure of tornado strength.

The limitation of the F-scale and EF-scale ratings as a measure of risk stems from reliance on damage to the built environment to assign ratings. Tornadoes in sparsely populated areas are less likely to strike buildings, and especially well-constructed structures as opposed to barns or outbuildings. Tornadoes can leave evidence of their destructive power in damage to the natural envi-ronment, like to trees or the ground; other indicators of powerful tornadoes include damage to asphalt roads, the size of debris, and the distance these pieces of debris are carried. Yet tornadoes are not typically rated stronger than

TABLE 2.15. Strong and Violent Tornadoes and Population Density

	F2+ Tornado Rate			Damage Area		
Density	.111** (9.87)	.286** (10.45)		4.28E−4** (2.67)	7.33E−4* (1.84)	
Density^2		−.00251** (6.99)			−4.38E-6 (0.84)	
Density10–49.9			1.05** (9.61)			.00816** (4.08)
Density50–99.9			1.54** (10.04)			.00926** (3.35)
Density100–499.9			1.91** (10.96)			.0130** (4.12)
Density500–999.9			3.51** (7.92)			.0187* (2.34)
Density1,000+			4.38** (10.84)			.0217** (3.05)

* and ** indicate significance from zero at the .10 and .01 levels in a two-tailed test, respectively.

F1 or F2 based only on damage to the natural environment; survey teams wish to see damage to buildings and particularly to well-built structures before rating tornadoes F3 or stronger. As a consequence, some powerful tornadoes that would cause F3 or stronger damage if they struck buildings are underrated in the F-scale, and this is continuing with the EF-scale. In essence, F-scale and EF-scale ratings are conservative and assign the minimum observed damage to buildings, not the likely strength of the tornado.

To explore the potential observed damage bias, we perform the same regressions of county tornado rate and damage area for the 12 tornado-prone states employed earlier using the F2 and stronger tornado rate and damage areas. Table 2.15 presents the three regression specifications for the F2+ tornado rate per 10,000 mi^2 and F2+ damage area. The same population density bias as for all tornadoes is observed. Population density significantly increases the F2+ tornado rate, with a positive and significant coefficient on the linear population density term. In the second column we see that the F2+ tornado rate increases but at a decreasing rate, as the population density squared term is negative and significant. Finally, with the density interval variables we can see more easily the magnitude of the underestimation. A county with a population density greater than 1,000 persons/mi^2 experiences about four more F2 or stronger tornadoes per 10,000 mi^2 than a county with a density of fewer than 10 persons/mi^2. To place this in perspective, the overall F2+ tornado rate for the entire state of Oklahoma is about 2.1 per 10,000 mi^2, the highest in the nation (Table 2.7). The frequency of powerful tornadoes appears to be substantially underestimated due to requiring damage to well-built structures in order to assign a higher F-scale rating. The last three columns of Table 2.15 indicate that the underestimation of powerful tornadoes extends to damage areas. The population density interval variables all attain statistical significance.

The minimum observed damage bias in F-scale ratings has two consequences for societal impacts research. First, tornadoes are more likely to be underrated in sparsely populated counties where there are few homes and other well-constructed buildings. This leads to a difference in the relative frequency of F2 and stronger tornadoes across counties or states, so rural areas will appear to have less risk of strong and violent tornadoes than urban areas. This impression, although incorrect, could nevertheless affect the siting of regional facilities like power plants. Second, total exposure of the nation to powerful tornadoes is being underestimated. The United States experiences about one F5 and ten F4 tornadoes per year, and about 70 mi^2 of violent tornado damage per year. Violent tornadoes are very rare, and so we must average events over a relatively large geographical area to estimate a rate. But the area over which we average will usually include sparsely populated counties, and some F4 or F5 tornadoes in these less populated areas will have been rated F1 or F2 instead. Thus the total incidence of violent tornadoes—which account for a disproportionate share of societal impacts—is underestimated. Investments in durable protection or in observing and prediction systems for tornadoes will be undervalued as a result.

The NWS essentially takes a conservative approach to assigning F-scale ratings to tornadoes: Tornadoes must prove themselves worthy of a strong or violent rating by damaging or destroying well-constructed buildings. In one way this is good, as it reduces the possibility of subjective overrating of tornadoes. But assigning a high F-scale rating only when a tornado destroys well-constructed buildings leads to the underrating of powerful tornadoes and creates a substantial bias in the frequency of strong and particularly violent tornadoes. As the most substantial societal impacts will result from strong and particularly violent tornadoes, societal risk is underestimated in an important way. Perhaps more significantly, relative risk across states or regions could be biased, with the apparent risk of powerful tornadoes in rural areas being substantially underestimated.

2.6. Summary

This chapter has examined patterns in tornado occurrence over time, across the country, and throughout the year, as these elements of climatology affect the societal impact of tornadoes. Although a sharp increase in reported tornadoes is apparent in the record, examination of numbers of reported strong and violent tornadoes or the total damage area reveals no increase over time. The best interpretation of the increase in total tornadoes over time is an increase in the likelihood that weak, short-lived tornadoes will be observed and documented and enter into official counts. The underlying tornado frequency in the United States does not appear to be changing over time. Several seasonal and regional patterns are apparent as well. Many states outside of Tornado Alley face considerable tornado risk, especially in the Southeast. Plains states face a well-defined tornado season in the late spring and summer months, while tornadoes occur more evenly throughout the year in the Southeast. Existing tornado records suffer from several biases that must be accounted for in evaluating the cost effectiveness of warnings or mitigation. Tornado records suffer from a population density bias, as tornadoes in sparsely populated areas are less likely to be reported and receive a lower rating on the Fujita scale because these storms are less likely to strike well-constructed buildings. Measures of tornado risk are likely to substantially underestimate risk in rural states, and the annual number of very powerful tornadoes is also likely underestimated as well.

AN ANALYSIS OF TORNADO CASUALTIES

3

3.1. Introduction

The National Weather Service (NWS) is charged with protecting lives and property from dangerous weather. Tornadoes are nature's most violent local storms, capable of flattening entire towns, so we will focus first on casualties. An examination of the determinants of fatalities and injuries provides value for two reasons. First, assessment of NWS efforts to reduce the lethality of tornadoes—issuing tornado warnings, improving the quality of warnings, and providing public education—requires analysis of the determinants of casualties. Consider the 1990s installation by the NWS of the NEXRAD network of Doppler weather radars (WSR-88D), at a cost of over $1 billion. The radars were expected to yield a significant societal benefit: improved tornado warnings, and a consequent reduction in casualties. Has the NEXRAD network delivered on this promise, and if so, how many fatalities have been prevented? A detailed analysis is required, because Doppler radar is just one of many factors affecting tornado casualties. Failure to control for other factors may leave us unable to identify the impact of the NWS efforts on casualties.

Second, analysis of tornado casualties can serve as a guide for further reducing casualties. An analysis of societal hazard impacts is akin to a medical diagnosis: Doctors must diagnose the ailment in order to successfully treat a patient. If patterns of casualties are like a patient's symptoms, an analysis of casualties is like the tests needed to confirm the diagnosis. Understanding tornado casualties can help avoid wasting resources on ineffective efforts to reduce casualties.

The analysis proceeds in this chapter as follows. Section 3.2 examines patterns in casualties: over time, by state, by Fujita Scale rating, and so forth—it's somewhat like describing a patient's symptoms. Section 3.3 sets the stage for our diagnostic tests by describing the variables we will include in our analysis. Section 3.4 presents the results of our tests, and multiple regression analysis of fatalities and injuries. (An appendix to this chapter includes the technical details of the models and the full regression results.) Section 3.5 provides a closer analysis of three sources of elevated casualties: tornadoes that strike at night, the winter season, and mobile-home vulnerabilities. All three vulnerabilities have a strong southeastern U.S. component. Section 3.6 considers state casualty rates, identifying states where tornadoes are more likely to kill or injure residents. We construct a casualty index that exhibits declining vulnerability along a southeast-to-northwest axis, with a peak in Florida and a minimum in North Dakota. The state casualty index is correlated with state forest cover, suggesting that an inability to see approaching tornadoes may be a significant determinant of casualties. Section 3.7 approaches the casualties question from a different perspective, considering how likely a person is to be killed or injured if a tornado strikes his or her residence, and offers estimates for both permanent and mobile homes. Section 3.8 explores potential fatalities in a worst-case scenario. Such an analysis can help emergency managers and the medical community prepare for disasters. We discuss what might qualify as a plausible "worst-case" scenario, and apply our analysis to estimate to this extreme possibility. The final section concludes.

3.2. Patterns of Tornado Casualties

We begin by considering patterns of fatalities and injuries over time. Figures 3.1 and 3.2 display annual fatality and injury totals in the United States from 1900 to 2007. A reduction in tornado lethality is readily apparent. The single deadliest year for tornadoes was 1925 with 797 deaths, while the highest injury total of 6,539 was in 1974. The fewest fatalities (15) occurred in 1986, and the smallest injury total (200) was in 1910. Not surprisingly, the biggest fatality and injury totals tend to occur together, with a correlation between

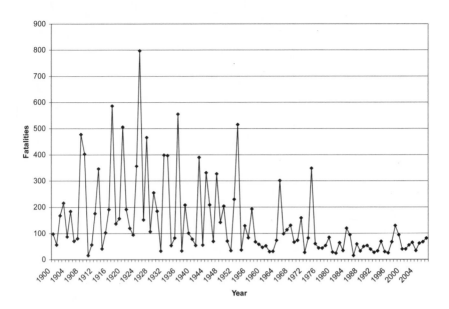

FIGURE 3.1. Fatalities by year, 1900–2007. Sources: Authors' calculations based on Grazulius and Storm Prediction Center archive.

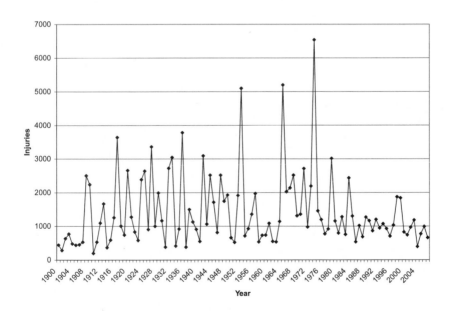

FIGURE 3.2. Injuries by year, 1900–2007. Sources: Authors' calculations based on Grazulius and Storm Prediction Center archive.

the annual totals of +.73. It's possible that the recorded totals, particularly for injuries, deviate from the true (but unobserved) totals due to errors in reporting, and such potential for error may be greater for earlier tornadoes, but despite this caveat, a downward trend does appear to exist, particularly for fatalities. Total fatalities in a decade fell below 1,000 for the first time in the 1960s, and for the last 30 years fatalities have averaged around 55 per year. Injuries in the 1980s and 1990s were about half of the total from the 1970s, and injuries this past decade have averaged just over 800 a year.

The reduction in casualty totals has occurred despite an increase in U.S. population from 76 million in 1900 to an estimated 302 million in 2007. To control for population change, Figures 3.3 and 3.4 display fatality and injury rates per million U.S. residents since 1900. The decline in casualties is even more pronounced when controlling for population growth. Year-to-year casualty totals exhibit considerable variation since the numbers of tornadoes, violent tornadoes, and powerful tornadoes striking populated areas vary each year, and this can obscure a decline in casualties. To smooth out the fluctuations, Figures 3.3 and 3.4 display a moving average of the annual rates. The moving average highlights the decline in fatalities, from a high of 2.9 per million in the mid-1920s (based on the 1925 Tri-State Tornado) to around .2 per million in recent years. The moving average for injuries declined from over 15 per million in the 1920s to under 4 per million in recent years. Note that scaling tornado casualties using U.S. population overstates the decline in the casualty rates because the states in Tornado Alley have grown more slowly than the United States as a whole. For instance, the population of eight states in the heart of Tornado Alley only doubled between 1900 and 2007, compared to the quadrupling of the U.S. population overall. The reduction of employment in agriculture has contributed to population declines in many counties in these states: In 1900, the population of Tornado Alley was more evenly distributed on account of labor-intensive agriculture, and as a consequence more people were likely to live in the path of a tornado striking an agricultural area.

Tornadoes do not occur uniformly throughout the year, and neither do tornado casualties. Figure 3.5 shows the percentage of fatalities and injuries by month. April accounts for the most fatalities and injuries, with nearly 30% of all casualties. May ranks second with 21% of fatalities and 18% of injuries, followed by March with 14% and 12%, and June with 11% of both fatalities and injuries. Almost three quarters of casualties occur in the spring months of March–June, consistent with the "tornado season." Each of the other months accounts for less than 8% of yearly casualties, with a secondary

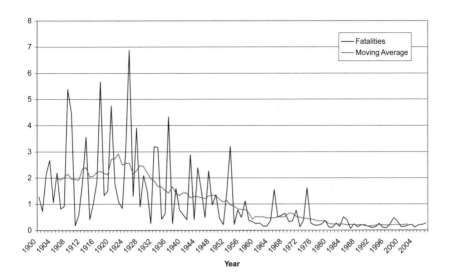

FIGURE 3.3. Fatalities per million persons. Sources: Authors' calculations based on Grazulius and Storm Prediction Center archive.

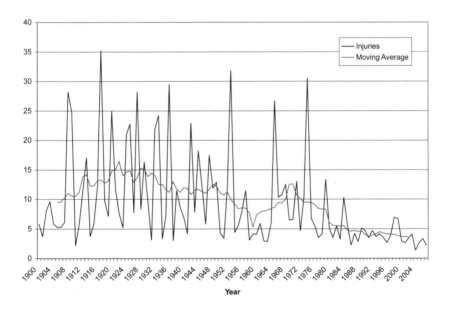

FIGURE 3.4. Injuries per Million Persons. Sources: Authors' calculations based on Grazulius and Storm Prediction Center archive.

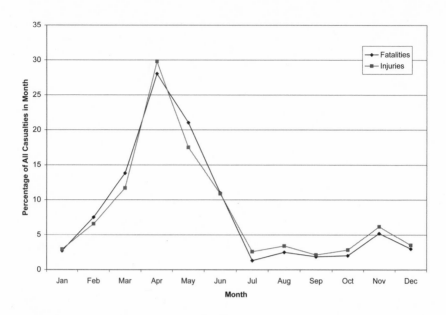

FIGURE 3.5. Fatalities and injuries by month

peak in November corresponding to the fall tornado season. Likewise, tornadoes are not equally dangerous throughout the year. Figures 3.6 and 3.7 report fatalities and injuries per tornado by month, respectively. Although most casualties occur during the tornado season between March and June, tornadoes in these months are not particularly dangerous. February is actually the most dangerous month for both fatalities per tornado (.29) and injuries per tornado (4.3). Tornadoes are particularly benign in the late summer, with the lowest fatalities and injuries per tornado occurring in July. The difference between February and July is quite substantial: The fatality rate in July (.01) is about 1/25 of the February rate, while the July injury rate (.4) is less than one-tenth of the February rate. Indeed, fatalities follow a distinct pattern within the year, with the lowest rates per tornado in the months from May to October and the highest rates in the months from November to April. These figures do not control for the strength of tornadoes, but as we saw in Chapter 2, most violent tornadoes occur in the spring, so this difference cuts against the timing of the strongest tornadoes.

Tornado casualties also vary considerably based on the time of day. To facilitate comparison, we break the day into five periods: overnight (12:00 to 5:59 AM), morning (6:00 to 11:59 AM), early afternoon (12:00 to 3:59 PM), late afternoon (4:00 to 7:59 PM), and evening (8:00 to 11:59 PM). The different parts of the day present different vulnerabilities: For instance, residents are

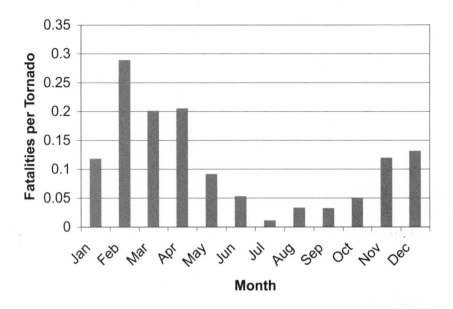

FIGURE 3.6. Fatalities per tornado by month

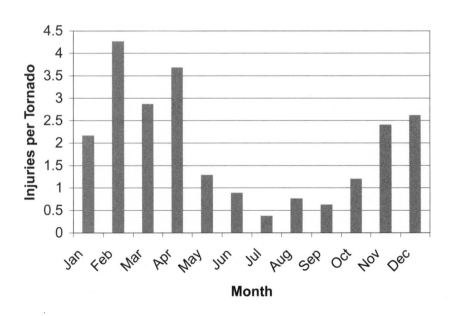

FIGURE 3.7. Injuries per tornado by month

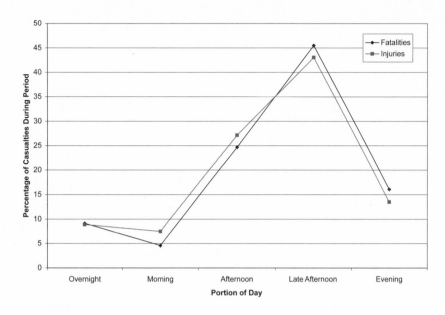

FIGURE 3.8. Casualties by time of day

likely to be at home and asleep during overnight hours, while the morning and late afternoon periods encompass rush hours. People are likely to be at work or school in the afternoons and at home in the evenings, but in either case will be awake and able to receive warnings. Most tornadoes occur in the late afternoon, as daytime heating fuels thunderstorms, and as Figure 3.8 shows, casualties are similarly distributed. About 45% of fatalities and injuries occur during the late afternoon and 25% in the early afternoon, followed by the evening, overnight, and morning hours. The dangerousness of tornadoes also seemingly varies across the day as well, as illustrated by Figures 3.9 and 3.10. Casualties per tornado rise during the day from the morning minimum to the overnight maximum: The overnight period exhibits the highest rates for both fatalities and injuries (.13 and 2.1 per tornado, respectively), while morning tornadoes are least dangerous (.05 and 1.3 per tornado). Fatalities vary more across the day, as fatalities per tornado overnight exceed the morning rates by a factor of three, while injuries per tornado overnight are about two-thirds greater. Residents are likely asleep during the overnight hours, which could explain this vulnerability, while the casualty rates in the late afternoon plausibly reflect the occurrence of the strongest tornadoes at these times.

We next consider the pattern of casualties across the country. Fatalities occurred in 41 states between 1950 and 2007, and injuries in every one of

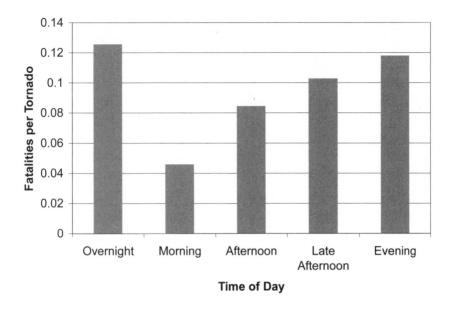

FIGURE 3.9. Fatalities per tornado by time of day

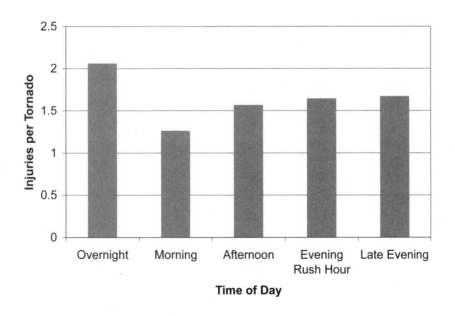

FIGURE 3.10. Injuries per tornado by time of day

the 48 contiguous states. Table 3.1 displays the fatality and injury totals by state. Texas had the most fatalities and injuries of any state, at 536 and 8,155 respectively. Mississippi, Alabama, and Arkansas rank second, third, and fourth in both fatalities and injuries, with Tennessee fifth in fatalities and Ohio fifth in injuries. Eighteen states suffered 100 or more fatalities, and 24 states experienced in excess of 1,000 injuries. Despite the popular association of risk with the Plains states of Tornado Alley, three of the top five states in fatalities were in the Southeast.

Casualty totals can be misleading, since they do not control for differences in population. Thus Table 3.1 also reports annual fatality and injury rates per million state residents from 1950 to 2007.[1] The U.S. fatality and injury rates over this period are 0.40 and 6.6, respectively, allowing us to see states with above- and below-average casualty rates. Tornado casualties are regionally concentrated in the central and southeastern United States: The states with the highest fatality rates are Mississippi at 2.9 deaths per million residents, followed by Arkansas at 2.7, Alabama and Kansas at 1.7, and Oklahoma at 1.6. These five states also have the highest injury rates but in a slightly different order: Mississippi at 40.4 per million, Arkansas at 37.6, Oklahoma at 25.4, Alabama at 25.4, and Kansas at 20.6. Texas illustrates the importance of controlling for state population, as it ranks only 10th and 14th in fatality and injury rates, respectively, despite having the most fatalities and injuries overall. The differences in casualty rates across the United States are substantial: For injury rates, the difference exceeds three orders of magnitude, as does the difference in fatality rates between Mississippi and New Jersey; and note that seven states experienced no fatalities at all.

Do the residents of the seven states with no fatalities since 1950 face no risk of death from tornadoes? Clearly, the answer to this question is no. Each of these states experienced tornadoes and injuries, and observed fatalities depend on the protective actions residents take. If residents take cover and no fatalities occur, it may still have been possible for fatalities to result if residents had ignored the risk. And the true, unobserved long-run fatality rates in these states, though likely close to zero, are still positive; 58 years has simply not been a long enough time to observe a fatality.[2] This likelihood—that the true fatality rate is not zero even in states that have not experienced a fatality—is relevant for the estimation of lives that could be saved with mitigation. The value of mitigation is not zero simply because no fatalities have occurred. Its valuation would require an estimate of the true fatality rate and the reduction in rate accordingly due to mitigation. Mitigation would be unlikely to save many lives in these states because the fatality rate would still be zero.

TABLE 3.1. Fatalities and Injuries by State, 1950–2007

State	Fatalities	Injuries	Fatality Rate	Injury Rate
Alabama	369 [3]	5,432 [3]	1.7230 [3]	25.364 [4]
Arizona	3 [35]	139 [34]	0.0202 [37]	0.937 [37]
Arkansas	337 [4]	4,703 [4]	2.6966 [2]	37.633 [2]
California	0 [t42]	87 [39]	0 [45]	0.067 [46]
Colorado	4 [t32]	194 [32]	0.0262 [34]	1.272 [34]
Connecticut	4 [t32]	700 [25]	0.0238 [35]	4.168 [27]
Delaware	2 [t36]	73 [39]	0.0617 [30]	2.250 [28]
Florida	160 [14]	3,319 [12]	0.3112 [23]	6.456 [22]
Georgia	170 [13]	3,639 [9]	0.5478 [14]	11.726 [12]
Idaho	0 [t42]	11 [45]	0 [45]	0.218 [44]
Illinois	202 [t10]	4,049 [7]	0.3206 [20]	6.426 [23]
Indiana	248 [7]	3,884 [8]	0.8301 [7]	13.001 [9]
Iowa	68 [22]	1,980 [17]	0.4182 [17]	12.177 [10]
Kansas	228 [9]	2,758 [14]	1.7014 [4]	20.581 [5]
Kentucky	116 [16]	2,967 [13]	0.5828 [13]	14.906 [6]
Louisiana	153 [15]	2,579 [15]	0.7041 [9]	11.869 [11]
Maine	1 [t39]	19 [44]	0.0159 [39]	0.302 [43]
Maryland	7 [29]	309 [29]	0.0306 [32]	1.351 [32]
Massachusetts	102 [17]	1,358 [21]	0.3138 [22]	4.177 [26]
Michigan	242 [8]	3,349 [11]	0.4854 [15]	6.718 [20]
Minnesota	94 [20]	1,862 [19]	0.4125 [18]	8.172 [16]
Mississippi	404 [2]	5,673 [2]	2.8797 [1]	40.437 [1]
Missouri	202 [10.5]	1,879 [18]	0.7311 [8]	6.801 [19]
Montana	2 [37]	23 [t42]	0.0465 [31]	0.535 [39]
Nebraska	54 [23]	1,166 [24]	0.6152 [12]	13.283 [8]
Nevada	0 [45]	2 [48]	0 [45]	0.042 [47]
New Hampshire	0 [45]	28 [41]	0 [45]	0.563 [38]
New Jersey	1 [40]	65 [40]	0.0025 [41]	0.162 [45]
New Mexico	5 [31]	155 [33]	0.0710 [29]	2.201 [29]
New York	21 [27]	307 [30]	0.0208 [36]	0.304 [42]
North Carolina	97 [19]	2,162 [16]	0.2929 [25]	6.528 [21]
North Dakota	24 [26]	317 [28]	0.6527 [11]	8.621 [15]
Ohio	184 [13]	4,393 [5]	0.3105 [24]	7.413 [17]
Oklahoma	265 [6]	4,115 [6]	1.6373 [5]	25.424 [3]
Oregon	0 [45]	3 [47]	0 [45]	0.022 [48]
Pennsylvania	83 [21]	1,235 [23]	0.1233 [27]	1.834 [30]
Rhode Island	0 [45]	23 [t42]	0 [45]	0.425 [40]
South Carolina	54 [24]	1,267 [22]	0.3154 [21]	7.400 [18]
South Dakota	18 [28]	461 [27]	0.4496 [16]	11.513 [13]
Tennessee	269 [5]	3,649 [9]	1.0727 [6]	14.551 [7]
Texas	536 [1]	8,155 [1]	0.6883 [10]	10.473 [14]
Utah	1 [40]	93 [37]	0.0128 [40]	1.194 [35]
Vermont	0 [45]	10 [46]	0 [45]	0.357 [41]
Virginia	27 [25]	540 [26]	0.0914 [28]	1.829 [31]
Washington	6 [30]	303 [31]	0.0264 [33]	1.332 [33]
West Virginia	2 [37]	103 [35]	0.0185 [38]	0.955 [36]
Wisconsin	100 [18]	1,599 [20]	0.3865 [19]	6.180 [24]
Wyoming	4 [31.5]	96 [36]	0.1745 [26]	4.189 [25]

The state casualty rates reported in Table 3.1 are based on almost 60 years of data, and may appear to reflect vulnerability reasonably well. But tornadoes are very low frequency events, and violent tornadoes only rarely strike a densely populated area. A longer time series consequently has value in estimating fatalities, particularly in states with lower tornado rates or lower population density. To explore how a longer time horizon affects assessment of relative state vulnerability, we construct state fatality totals for the period 1900–1949 using Grazulis's tornado records (1993). As we extend the time series into the past, the potential for error in the totals increases; Grazulis constructed his archive based on contemporary reports, which are sometimes ambiguous, especially regarding the potential double counting of fatalities. Despite the potential for error, the value of an additional 50 years of data to assess relative casualty rates makes the trade-off worthwhile. We again calculate fatalities per year per million residents, with average state population taken from the decennial censuses from 1900 through 1950.

Table 3.2 presents the fatality totals and rates for the period 1900–1949, along with the totals and rates for 1900–2007, with state ranks in parentheses. The fatality totals and rates for 1900–1949 are much higher, reflecting the greater lethality of earlier tornadoes identified by Brooks and Doswell (2002). Comparisons of the pre-1950 and post-1950 state fatality rates are not very revealing. Of greater interest is the rank of the states, and whether a state experienced a substantial change in rank between 1900–1949 and 1950–2007, which can be determined by comparing columns 2 and 4 of Table 3.2 with the ranks reported in Table 3.1. The rank of most states remains fairly consistent; the Spearman Rank Correlations for the fatality totals and rates for 1900–1949 and 1950–2007 are +.88 and +.89, respectively. A high-risk state since 1950 was usually a high-risk state before 1950. For instance, Arkansas and Mississippi rank 1st and 2nd in fatality rate over each period, simply reversing order, while Alabama and Oklahoma are also in the top five in fatality rate in each period. Five of the seven states with no fatalities since 1950—California, Nevada, Oregon, Rhode Island, and Vermont—did not experience a fatality between 1900 and 1949 either. Several states, however, did move five or more spots in the fatality rate rankings; most of these cases reflect the influence of one major killer tornado in a state where fatalities are rare. For example, Massachusetts ranked 22nd in fatality rate over the 1950–2007 period, due to the 90 fatalities in the June 1953 tornado in Worcester, but experienced only 4 fatalities and ranked 37th in fatality rate for 1900–1949. Michigan ranked 15th in fatality rate over the 1950–2007 period, but had only 33 fatalities and ranked 29th in fatality rate for 1900–1949. West Virginia had 2 fatalities between 1950 and 2007

TABLE 3.2. State Tornado Fatalities, 1900–1949 and 1900–2007

State	Fatalities, 1900–1949	Fatality Rate 1900–1949	Fatalities 1900–2007	Fatality Rate 1900–2007
Alabama	977 [5]	7.892 [4]	1346 [4]	3.671 [4]
Arizona	0 [43.5]	0 [46]	3 [38.5]	0.016 [38]
Arkansas	1048 [2]	12.149 [1]	1385 [3]	6.001 [1]
California	0 [43.5]	0 [43.5]	0 [46]	0 [46]
Colorado	30 [28]	0.625 [22]	34 [30]	0.156 [28]
Connecticut	0 [43.5]	0 [43.5]	4 [37]	0.015 [39]
Delaware	1 [37]	0.084 [34]	3 [38.5]	0.062 [34]
Florida	31 [27]	0.444 [24]	191 [21]	0.301 [24]
Georgia	733 [7]	5.115 [6]	903 [7]	1.823 [8]
Idaho	2 [36]	0.097 [33]	2 [40]	0.026 [35]
Illinois	993 [4]	2.893 [12]	1195 [6]	1.133 [12]
Indiana	277 [11]	1.773 [16]	525 [11]	1.063 [14]
Iowa	127 [19]	1.052 [21]	195 [20]	0.629 [19]
Kansas	271 [13]	3.083 [11]	499 [12]	2.053 [5]
Kentucky	189 [16]	1.486 [17]	305 [16]	0.858 [16]
Louisiana	515 [9]	5.156 [5]	668 [10]	1.946 [7]
Maine	0 [43.5]	0 [43.5]	1 [42]	0.009 [42]
Maryland	30 [29]	0.370 [26]	37 [29]	0.110 [31]
Massachusetts	4 [32]	0.021 [37]	106 [25]	0.188 [27]
Michigan	33 [26]	0.156 [29]	275 [17]	0.361 [22]
Minnesota	218 [14]	1.798 [15]	312 [15]	0.822 [17]
Mississippi	1061 [1]	11.060 [2]	1465 [2]	5.689 [2]
Missouri	559 [8]	3.168 [10]	761 [8]	1.539 [10]
Montana	3 [35]	0.126 [32]	5 [35.5]	0.069 [33]
Nebraska	274 [12]	4.341 [7]	328 [14]	1.981 [6]
Nevada	0 [43.5]	0 [43.5]	0 [46]	0 [46]
New Hampshire	1 [38]	0.043 [35]	1 [42]	0.013 [40]
New Jersey	4 [31]	0.0233 [36]	5 [35.5]	0.008 [43]
New Mexico	3 [34]	0.143 [31]	8 [32]	0.081 [32]
New York	9 [30]	0.016 [38]	30 [31]	0.018 [37]
North Carolina	57 [22]	0.392 [25]	154 [22]	0.299 [25]
North Dakota	55 [23]	1.893 [14]	79 [26]	1.097 [13]
Ohio	152 [17]	0.504 [23]	336 [13]	0.347 [23]
Oklahoma	956 [6]	10.027 [3]	1221 [5]	4.356 [3]
Oregon	0 [43.5]	0 [43.5]	0 [46]	0 [46]
Pennsylvania	67 [21]	0.153 [30]	150 [23]	0.124 [30]
Rhode Island	0 [43.4]	0 [43.5]	0 [46]	0 [46]
South Carolina	208 [15]	2.424 [13]	262 [18]	0.937 [15]
South Dakota	39 [24]	1.287 [19]	57 [28]	0.741 [18]
Tennessee	478 [10]	3.733 [9]	747 [9]	1.819 [9]
Texas	1032 [3]	3.923 [8]	1568 [1]	1.391 [11]
Utah	0 [43.4]	0 [43.5]	1 [42]	0.009 [41]
Vermont	0 [43.5]	0 [43.5]	0 [46]	0 [46]
Virginia	39 [25]	0.320 [27]	66 [27]	0.146 [29]
Washington	0 [43.5]	0 [43.5]	6 [34]	0.0186 [36]
West Virginia	112 [20]	1.448 [18]	114 [24]	0.573 [21]
Wisconsin	148 [18]	1.073 [20]	248 [19]	0.576 [20]
Wyoming	3 [33]	0.300 [28]	7 [33]	0.198 [26]

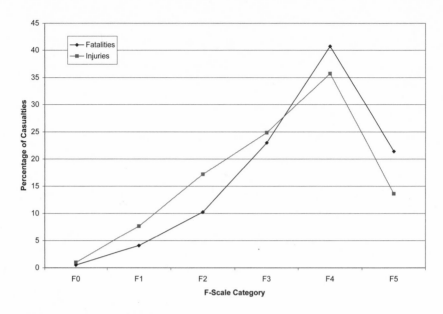

FIGURE 3.11. Casualties by F-scale

and ranked 38th in fatality rate, but because of a particularly deadly tornado outbreak in 1944, had 112 fatalities and ranked 18th for 1900–1949; therefore, it would be classified as a low fatality risk state over the past 60 years, and yet based on the entire century faces a modest risk. Nebraska similarly experienced relatively more fatalities in the early 20th century, ranking 7th in fatality rate for 1900–1949 compared with 12th in the more recent decades. Illinois, Georgia, and South Carolina ranked lower in fatality rate over the period 1950–2007 than over the entire century, while Kansas has gone from 11th in fatality rate in the first half of the century to 4th since 1950.

Damage as measured by the rating of a tornado on the Fujita Scale affects casualties. Figure 3.11 shows the percentage of fatalities and injuries accounted for by tornadoes in each F-scale category for the period 1950–2007.[3] Casualties result mainly from strong (F2 and F3) and violent (F4 and F5) tornadoes. The 1.2% of tornadoes rated F4 or F5 account for 62% of fatalities and 49% of injuries. By contrast, 78% of tornadoes were weak (rated F0 and F1), and account for only 5% of fatalities and 9% of injuries. The difference in lethality of tornadoes across the F-scale categories is better reflected in fatalities and injuries per tornado, displayed in Figures 3.12 and 3.13. More damaging tornadoes are clearly more dangerous: Fatalities per tornado differ by a factor of 15,000 when comparing F5 tornadoes (16.3) with F0 tornadoes (.001), while injuries per tornado differ by a factor of 5,000.

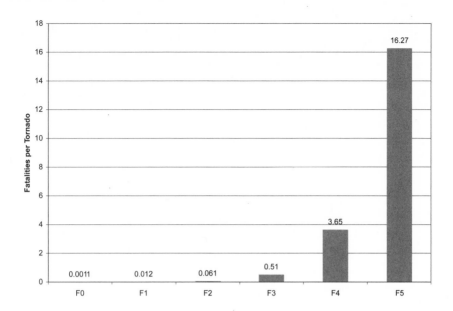

FIGURE 3.12. Fatalities per tornado by F-scale category

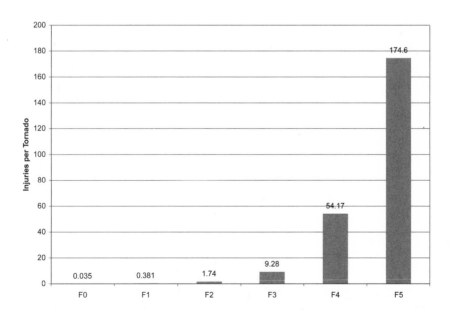

FIGURE 3.13. Injuries per tornado by F-scale category

As the distribution of casualties by F-scale suggests, a handful of tornadoes result in the overwhelming majority of casualties, while most tornadoes are not killers. Of the more than 50,000 state tornadoes that occurred since 1950, only 1,295, or 2.6%, killed one or more persons, and only 13% resulted in one or more injuries. And casualties were concentrated even within the casualty-producing storms: 1% of tornadoes between 1950 and 2007 (about 40% of killer tornadoes and 7% of injury tornadoes) accounted for 82% of fatalities and 66% of injuries. A total of 102 and 153 tornadoes killed 10 or more persons or injured 100 or more, and these tornadoes (about .2% and .3% of all tornadoes over the period) accounted for 49% of fatalities and 45% of injuries. Fatalities are even more concentrated than injuries, suggesting that fatalities might depend on factors that are hard to control for, or what might appear as randomness. The majority of tornado casualties stem from a small number of very dangerous storms, and the way that society addresses these storms (through the warning and response process) will dramatically affect casualties over the long term.

The clustering of casualties across time is even greater because strong and violent tornadoes tend to occur in major outbreaks. Killer tornado days thus provide another perspective on the clustering of fatalities. The 1,295 killer tornadoes between 1950 and 2007 occurred on 795 different days. Table 3.3 lists the 10 deadliest tornado days since 1950. The deadliest day was April 3, 1974, the day of perhaps the most extensive and powerful tornado outbreak in U.S. history, with 292 fatalities from 47 killer tornadoes across 11 states. The April 11, 1965 (Palm Sunday) outbreak ranks second, with 260 fatalities from 23 killer tornadoes across five states. The fourth and eighth deadliest days occurred on June 8 and 9, 1953, with an F5 tornado in Flint, Michigan, on June 8 and an F4 tornado in Worcester, Massachusetts, the next day. Tornado fatalities occurred on less than 4% of days between 1950 and 2007, and the 10 days in Table 3.3 alone account for 30% of tornado fatalities over the period. Only one of the deadliest days resulted from a single killer tornado, demonstrating how the most deadly tornadoes occur in powerful outbreaks. The prevalence of super tornado outbreaks implies that accurately assessing the impact of new warnings on casualties, for example, cannot occur until the next major outbreak.

It's common knowledge that mobile or manufactured homes are vulnerable to tornadoes; this phenomenon was even immortalized in an episode of television's *The Simpsons*. Researchers who previously examined the tornado–mobile-home problem include Golden and Adams (2000) and Golden and Snow (1991). The NWS has reported the location of tornado

TABLE 3.3. The 10 Deadliest Tornado Days, 1950–2007

Day	Fatalities	Killer Tornadoes	States
April 3–4, 1974	292	47	Alabama, Georgia, Illinois, Indiana, Kentucky, Michigan, North Carolina, Ohio, Tennessee, Virginia, West Virginia
April 11, 1965	260	23	Illinois, Indiana, Michigan, Ohio, Wisconsin
March 21, 1952	209	22	Alabama, Arkansas, Mississippi, Missouri, Tennessee
June 8, 1953	142	5	Michigan, Ohio
May 11, 1953	127	2	Texas
February 21, 1971	121	5	Louisiana, Mississippi
May 25, 1955	102	3	Kansas, Oklahoma
June 9, 1953	90	1	Massachusetts
May 31, 1985	76	11	Ohio, Pennsylvania
May 15, 1968	72	8	Arkansas, Illinois, Indiana, Iowa

Proportions of tornadoes occurring in each time period rated in each F-scale category

fatalities since 1985, as displayed in Figure 3.14. More than 43% of tornado fatalities occurred in mobile homes, followed by 31% in what the NWS describes as permanent homes. Other locations for fatalities include 9% in vehicles, and about 5% each in businesses, schools and churches, and outdoor locations. The occurrence of more fatalities in mobile homes than in permanent homes is remarkable, because mobile homes comprised only 7.6% of U.S. housing units in 2000, and only 6.9% of the population lived in mobile homes. As a result, the likelihood of being killed in a tornado is much greater for residents of mobile homes than of single-family homes; a simple calculation based on the numbers of mobile and permanent homes in 2000 suggests that fatalities are 15 times more likely in mobile homes.[4] (This is a rough calculation and does not control for the proportion of mobile homes in areas struck by tornadoes. We will return to fatality rates for mobile and permanent homes in Section 3.7 and Chapter 5.) The mobile-home problem is indeed real, and the vulnerability of mobile homes will have to be addressed to reduce tornado fatalities in the future.

NWS records on tornado fatalities allow us to analyze the age and sex of victims from 1996 to 2007. Men were slightly more likely to be killed than women, at 51.2% of fatalities, even though women comprise almost 51% of the U.S. population; still, fatalities are reasonably balanced between the sexes. Tornado victims tend to be older than the U.S. population as a whole, with

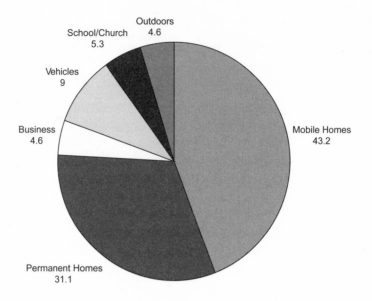

FIGURE 3.14. Fatalities by location, 1985–2007. Numbers are the percentage of fatalities in each location, as compiled by the National Weather Service.

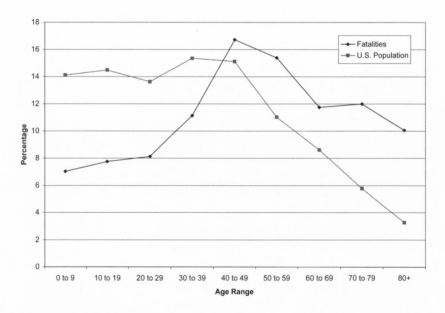

FIGURE 3.15. Age distribution of tornado fatalities and U.S. population. Calculations based on U.S. tornado fatalities, 1996–2007.

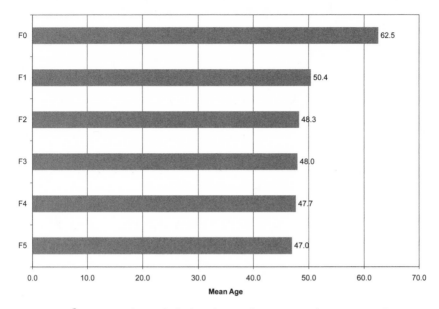

FIGURE 3.16. Mean age of tornado fatalities by F-scale rating. Based on U.S. tornado fatalities, 1996–2007.

an average age of 48.1 years, compared to a median age of the U.S. population of 35.3 years in 2000. Figure 3.15 presents the age distribution of tornado victims and the U.S. population in 2000. The percentage of victims exceeds the percentage of the population for the 50–59 years age group and all older cohorts. In particular, tornado victims are disproportionately over 70 years old; 19% of victims, but only 3.3% of the population, is over 80 years of age, and 12% of victims and less than 6% of the population is 70–79 years old. The percentage of fatalities was less than the percentage of the population for all age groups up to 40–49 years, with the percentage of victims under 20 about half of the proportion of the population under 20. One plausible explanation for the disproportional share of elderly victims might be a greater difficulty in hearing and quickly responding to tornado warnings.

We also calculated the average age of tornado fatalities by F-scale rating, time of day, and location. Figure 3.16 displays the mean age of victims by F-scale rating of the tornado. An inverse relationship exists between mean age and F-scale: Victims were older for lower rated tornadoes, at 62.5 years for F0 tornadoes and 50.4 years for F1 tornadoes (the number of F0 fatalities is quite small). The average age of victims for F5 tornadoes, 47.0 years, still exceeds the median age of the population by almost 12 years. Figure 3.16 suggests that more vulnerable individuals face greater relative risk from less

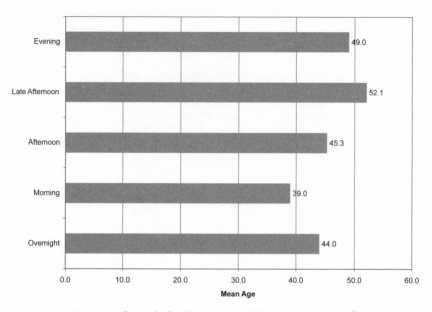

FIGURE 3.17. Mean age of tornado fatalities by time of day. Based on U.S. tornado fatalities, 1996–2007.

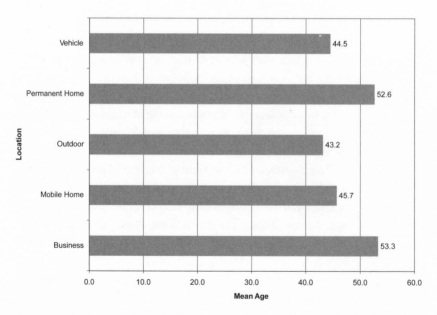

FIGURE 3.18. Mean age of tornado fatalities by location of fatality. Based on U.S. tornado fatalities, 1996–2007.

violent tornadoes. Figure 3.17 displays the mean age of victims by time of day of the tornado. Victims in the morning were the youngest on average, at 39.0 years, and the oldest victims on average were in the late afternoon, at 52.1 years. The average age of overnight victims, 44.0 years, suggests that when residents are caught by surprise in the middle of the night, young and old are both at risk, but in the late afternoon when it may be possible to see an approaching tornado, the elderly face a higher relative risk. Finally, Figure 3.18 displays the mean age of victims by location, for the locations with a sufficiently large number of fatalities for the mean to convey some significance. Permanent homes and businesses have the oldest victims on average (around 53 years of age), while the mean ages for victims in outdoor locations, vehicles, and mobile homes are all around 45 years. Vehicles, the outdoors, and mobile homes have greater vulnerability to tornadoes, while permanent homes and businesses likely have areas of relative safety. The pattern is again consistent with the vulnerable elderly facing a higher relative risk in less dangerous tornadoes, locations, and times. An inability to hear the approaching tornado or to move quickly to a safe location is more relevant when in a building that offers areas of relative safety.

3.3. Determinants of Tornado Casualties

Tornado fatalities and injuries have declined over time, as Figures 3.1 and 3.2 illustrate. Can we validly conclude that tornadoes have become less deadly over time? A satisfactory answer would provide statistical evidence supporting a decline in fatalities and injuries. That is, we would like to be able to reject a null hypothesis of no change in fatalities or injuries over time.

One way to approach this question is to examine the frequency of large-casualty tornadoes. Have mass-casualty events (i.e., tornadoes exceeding some threshold of casualties) become less frequent over time? Since 1900, 11 tornadoes have killed 100 or more persons in the United States, with the last occurring in 1953; thus, between 1900 and 1953, there were .204 100+ fatality tornadoes per year. If the annual probability of a 100+ fatality tornado of .204 has not declined, what is the likelihood of observing no 100+ fatality tornadoes in 58 years? The probability of no 100+ fatality tornadoes in a given year is .796, and so the probability of no 100+ fatality tornadoes in 58 successive years is $3.62*10^{-6}$, or very, very unlikely. We can conclude with statistical confidence that the probability of a 100+ fatality tornado has declined. The choice of this threshold might seem arbitrary, and so we can also consider 50+ fatality tornadoes. Thirty-one tornadoes since 1900 killed 50 or more persons, with the most recent in 1971. If the annual probability of a 50+ fatality tornado from 1900 to 1971 of .431 has not changed, the probability of no 50+ fatality tornadoes in 36 years is $8.9*10^{-10}$. Again, we can conclude with statistical confidence that the probability of a 50+ fatality tornado has declined. The same conclusion holds for 500+ injury tornadoes.

We could also fit a time trend to the casualty totals in Figures 3.1 and 3.2, or to casualties per million residents in Figures 3.3 and 3.4, to test if tornadoes have become less dangerous over time. We have done this using a simple linear regression with a constant and a time variable. Table 3.4 presents the estimated coefficient on the time trend variable, along with the standard error and p-value in a two-tailed test of the null hypothesis that the time trend coefficient is zero. We see that a statistically significant downward time trend exists for fatalities, fatalities per million residents, and injuries per million residents, but not for total injuries.

Annual casualty totals provide only weak evidence of a downward trend, however. Fatalities and injuries in a given year depend on many factors, including the total number of tornadoes and especially the number of strong or violent tornadoes. Based on this, we might want to scale casualty totals by the number of tornadoes and F2+ tornadoes in addition to population, say, fatalities per million residents per tornado. But even this is unlikely to be sufficient. Casualties also depend on the timing during the year and day as well, as illustrated in Figures 3.5 through 3.10, and the population in the path of each tornado. The decline in fatalities might be a result of any number of factors, as opposed to a reduction in the potential for a given tornado to kill residents.

Analysis of annual casualty totals would require an assumption that the underlying distribution of tornadoes, or more specifically the tornado threat (the number, timing, and strength of tornadoes), has not changed over time. A long time horizon for analysis will allow some of the year-to-year variation in tornado vulnerability to even out. For instance, F5 tornadoes do not occur every year, and when they do, they sometimes strike rural areas; however, over enough years, we will see powerful tornadoes hit populated areas. An unchanged tornado threat is a big assumption and questionable, given the low frequency of tornadoes. Consider tornado fatalities in Massachusetts relative to other New England states. The 58-year fatality rate in Massachusetts is .31 per million (Table 3.1), while the bordering states have a combined rate of .02 per million per year. Ninety of the 102 fatalities in Massachusetts are due to the F4 tornado that hit Worcester on June 9, 1953. The underlying

TABLE 3.4. Evidence on Time Trend in Tornado Casualties

Measure	Point Estimate	Standard Error	p-Value
Fatalities	−1.96	.414	.0000
Fatalities per Million	−.0242	.00321	.0000
Injuries	.231	3.47	.9471
Injuries per Million	−.0892	.0235	.0002

Results of a linear regression of the designated casualty measure on a constant plus time trend over the 1900–2007 period. The p-value is the probability that we would observe the calculated point estimate of the time trend if the true time trend were zero.

tornado rate in Massachusetts is similar to the surrounding states (Chapter 2), but even over almost 60 years one tornado still dominates any casualty comparison. It seems plausible that Massachusetts has similar vulnerability to tornado casualties as Connecticut or Rhode Island, but it may be another century before we could conclude this based on observed fatalities. In addition, climate changes over time, and there is no guarantee that over the next century the frequency of tornadoes or their geographic distribution will not change.

We believe that adjusting the annual totals for all of these various factors—the number, strength, and timing of tornadoes, the population of storm paths—is a dead end, even though a number of papers have indeed analyzed annual totals for natural hazards. A much better approach is to estimate a regression model of the number of persons killed or injured using the tornado as the unit of analysis. Variables can then be included to control for as many observable and measurable factors potentially affecting casualties as possible: the F-scale rating, the time of day, month of year, and so forth, as well as a variable for the year of the storm. Typically, data limitations prevent social scientists from conducting an ideal analysis. Nonetheless, regression analysis using the tornado as a data point allows a better way to examine whether tornadoes have become less deadly; specifically, we can test whether a year variable has a negative and statistically significant impact on casualties in a regression analysis. We will address this question in this chapter. In this section, we describe the variables included in the regressions.

Our regression models include many of the factors discussed in Section 3.2, including F-scale, month, time of day, as well as the effect of a tornado occurring on a weekend. To control for F-scale, we construct a set of dichotomous (dummy) variables for the different ratings, or F0, F1, F2, F3, F4, and F5. The variable F2, for instance, equals 1 for tornadoes rated F2 on the F-scale and 0 for all other tornadoes. Defining dummy variables for each category, instead of an integer variable ranging from 0 to 5, allows the impact of F-scale on casualties to vary across the scale. We would expect based on Figure 3.11 that a tornado rated higher on the F-scale would cause more casualties, but the dummy variables might reveal that F1 tornadoes are only somewhat more deadly than F0 tornadoes, while F4 tornadoes are substantially more deadly than F3 tornadoes. Three caveats are in order here. First, we are using the F-scale rating as a measure of tornado strength, although in truth it is a damage scale (Doswell and Burgess 1988). Ideally, we would like to have a measure of the wind speed in the tornado, but this is not available. In most cases this distinction will not matter much, as powerful tornadoes also do more damage and are more dangerous. But as discussed in Chapter 2, a tornado that does not strike well-constructed buildings may be rated F2 or less, even though it might have been capable of producing F4 or F5 damage if it struck a more populated area. Such a tornado is less likely to result in casualties because it probably did not strike many homes, but it might have encountered people in the open or in vehicles. A second potential limitation of the F-scale rating is the strengthening and weakening of tornadoes over their paths, as the rating is based on the maximum damage along the path. Different tornadoes might attain the same maximum, but differ substantially in average damage. For example, two tornadoes with 20-mile paths might both reach F4 strength, one for 100 yards along its path, and the other for 15 miles—although both tornadoes will be rated F4, the potential lethality of the second tornado is substantially greater. The third potential problem arises if casualties affect a tornado's F-scale rating. NWS personnel survey the tornado damage path to assign a rating, yet damage assessment has a subjective element. As discussed in Chapter 2, the Enhanced Fujita Scale is intended to reduce this subjectivity, although ratings may still be influenced by whether the tornado resulted in fatalities or injuries. Unfortunately, there is little we can do to correct for any of these potential problems, since we need a very large set of tornadoes for statistical analysis of the determinants of casualties, and the F-scale rating is the only variable available to control for the strength of a tornado.

We include dummy variables for the parts of the day examined in Section 3.2—night, morning, early afternoon, late afternoon, and evening—where the variable for each category equals 1 for a tornado during that time and 0 otherwise. We also construct month dummy variables to test for differences in casualties across the year: The variable May, for instance, equals 1 for May tornadoes and 0 otherwise. In addition, we include a dummy variable for tornadoes occurring on a weekend, since differences in daily routines on weekdays and weekends could affect tornado casualties. For instance, evening rush hour coincides with the peak time for tornadoes and particularly powerful tornadoes, but highways that would be crowded on a workday might be sparsely traveled on a Saturday evening. Likewise, people who would be at work or school on a weekday will be at home or engaging in outdoor recreation on the weekend. Some of these factors suggest that weekday tornadoes might be more dangerous (due to evening commutes) and others might make weekends more dangerous (residents at home in mobile homes). We will let the analysis show how the factors balance out.

We also include the path length in miles as a tornado characteristic variable. A tornado that is on the ground longer can potentially strike more homes, businesses, and vehicles. Comparison of the longest- and shortest-track tornadoes shows this factor's influence on casualties. The 5,000 tornadoes with the longest paths (about 10% of the tornadoes over the period) resulted in almost 3,500 fatalities and over 52,000 injuries, while the 5,000 shortest-path tornadoes killed 14 and injured 245. Stronger tornadoes tend to have longer paths (see Table 2.2), so some of this effect is due to F-scale, but path length itself clearly affects casualty potential.

We include a number of tornado-path characteristic variables. The hazard to humans posed by a tornado depends on the nature of the tornado and the characteristics of the damage path. We construct variables using the demographic and economic variables for the county or counties struck by the tornado. Counties are large relative to tornado damage-path areas; the average county struck by a tornado is about 1,000 mi^2, while an average tornado damage area is less than 1 mi^2. A county can include urban areas with population densities in excess of 1,000 persons per mi^2 and rural areas with few inhabitants. A more refined and accurate description of tornado paths would allow construction of variables more accurately characterizing the actual path, and in a perfect world, a digital map could generate an exact description of the tornado path and include the number of homes, mobile homes, and businesses in the path. Unfortunately, the most detailed reliable path description variable we have for the SPC tornado archive is the county.[5]

A reader could certainly question if county-level variables are precise enough to control for storm path characteristics. Social scientists often encounter such problems in empirical work, where the specific data needed to test predictions are simply not available. In such cases, researchers have no alternative to attempting the analysis with the available data. We were initially skeptical about whether the county variables could adequately control for storm-path characteristics, but the variables have performed well in explaining tornado casualties. If many control variables failed to attain significance or had unexpected signs, it could have signaled the inadequacy of county-level variables.

We include two path variables in the 1950–2007 analysis, population density and median family income. Population density controls for the number of persons in the path of a tornado. An increase in population should increase fatalities and injuries, everything else being equal. While imperfect, population density allows us to distinguish highly populated urban and sparsely populated rural counties. The population of the United States has grown considerably since 1950, but population growth accounts for only a small change in density relative to the difference between urban and rural counties. Granted, tornadoes will sometimes strike populated areas of rural counties and sparsely populated areas of urban counties, but with thousands of tornadoes, this should balance out. The inclusion of income as a control variable might surprise some readers: After all, nature does not care if people are rich or poor, so one might expect that wealthy persons or communities are just as vulnerable to tornado casualties as the poor. But economists have found that people consider safety to be a luxury good; that is, as people become wealthier, they spend an increasing share of their income to make their lives safer, through health care, safer cars, safer homes, and safer workplaces. In terms of tornadoes, higher incomes might lead residents to purchase NOAA Weather Radios or another emergency alert product, or to install tornado shelters in their homes. Wealth could also have community-level effects: wealthier communities might have tornado sirens or better emergency management and medical care. Research on natural hazards has found evidence of this "wealth effect" for hazards safety as well, and so we expect higher income to reduce casualties.[6] Income exhibits variation both across counties (poor versus wealthy communities) and over time, as the nation has become wealthier since 1950.

We include a number of other economic and demographic path characteristic variables in the 1986–2007 data set. One such variable is the proportion of county residents living in rural areas as defined by the Census Bureau. Rural areas tend to have lower population density, so this supplements population density in controlling for county population. For a given population density, a higher urban population likely means the county has more densely populated areas with the potential for more casualties, but also more open space where a tornado might result in few casualties. It is unclear whether an urban concentration of population will lead to more or fewer casualties.

We include one variable related to income, the proportion of county residents with income below the poverty line. The distribution of income, in addition to median income, has been identified as a significant determinant of natural hazard fatalities (Anbarci et al. 2005; Kahn 2005). Conventional wisdom holds that low-income households are more vulnerable to natural hazards, for the reasons discussed previously. If so, poverty might better explain casualties than median income, perhaps because a threshold exists in the relationship between hazards and vulnerability. Conceivably, households not in poverty might be able to purchase a level of tornado safety (a newer mobile home or a means to receive warnings in real time) that reduces vulnerability. Once this level is attained, higher income may have little impact on safety. If poverty contributes to tornado vulnerability, the poverty rate will increase tornado casualties.

We include three variables describing the built residential environment. We have already discussed the vulnerability of mobile homes to tornadoes, and thus we include mobile homes as a proportion of county housing units as a control variable. An increase in the proportion of mobile homes in a county should increase casualties, everything else being equal.[7] We also include the age of the median house as a control variable. The age is found by taking the difference between the year the tornado occurred and the median age of homes in the county. Progress in construction leads us to expect that newer homes will offer more protection against hazards. For instance, technology is continually making cars safer, while the Department of Housing and Urban Development implemented a construction code for manufactured housing in 1976. If these considerations hold, older homes will be associated with increased casualties.

Commuting patterns could affect vulnerability to tornadoes. Vehicles are vulnerable to tornadic winds, and the NWS recommends that occupants abandon their vehicles and lie in a ditch if a tornado approaches. We include the proportion of employed county residents who commute in excess of 30 minutes to control for commuting patterns. If long commutes increase vulnerability, this variable will positively affect casualties. Note that this variable also controls for residential patterns, as a county with a large proportion of long commutes is likely a suburban bedroom community.

Education may also affect vulnerability to hazards. More educated residents may be more likely to understand the difference between a tornado watch and a tornado warning, to access the Internet for weather information, and to otherwise appreciate the potential danger posed by tornadoes. We include two education variables, the proportion of county residents aged 25 or older who have a four-year college degree, and the proportion of residents 25 or older without a high school degree (or equivalent). These two variables allow investigation of whether high educational attainment decreases casualties, low educational attainment increases casualties, or both. We might find, for instance, that low educational attainment increases vulnerability, but that higher education does not further reduce casualties. Note that education, particularly a college degree, is correlated with income, and thus the college education variable may capture some of the impact of income on casualties.

We include four other demographic control variables. Age can affect vulnerability, as discussed above, so we include the proportion of county residents who are under 18 and over 65 years of age. The elderly are thought to be particularly vulnerable to hazards. As we saw in Section 3.2, the average age of tornado victims exceeds the median age of the population. For tornadoes, the relevant vulnerabilities include difficulties hearing sirens and taking cover quickly. Young persons may not appreciate the dangers posed by tornadoes, but if schools teach tornado safety, they may in fact be more knowledgeable about these dangers. Minority residents are often considered more vulnerable to natural hazards, and thus we include the proportion of nonwhite county residents as a control variable. Finally, the proportion of male county residents controls for any gender-based differences in attitudes about tornado risk.

3.4. Tornado Casualties: Evidence from Casualty Regressions

We conducted two different regression analyses of tornado casualties, one using tornadoes from 1950 to 2007, to allow for examination of long-term time trends, and a second using tornadoes from 1986 to 2007, to control for more economic and demographic variables. The 1986–2007 data set will also be used in our analysis of tornado warnings, as 1986 is the first year for which tornado warning verification records are available. (The appendix to this chapter will discuss the details of the regression models, present the raw regression results, and discuss the statistical significance of the results.) We have already examined the impact of several of the variables on casualties in Section 3.2, but those comparisons were univariate, meaning that they did not control for other factors that affect casualties. The univariate comparisons, of (for example) fatalities per tornado by month, provide expectations concerning the likely impact of variables on casualties. But the different variables might partially correlate with each other, and only a multivariate analysis can determine if the months of the year are significant when controlling for F-scale, time of day, and other factors. Although the regression results largely confirm the univariate comparisons, the analysis is quite different.

Investigation of long-term time trends in casualties would be facilitated if we could extend the data set back before 1950. To do this, we would need to combine the SPC data set with Grazulis's archive (1993). However, since Grazulis's records, by his own description, include only "significant" tornadoes, combining these two sets of records into one data set would cause a potential problem, as the inclusion of weaker tornadoes since 1950 would bias the analysis toward a decline in casualties over time. Even including the F-scale rating of the tornado may not adequately control the selection effect in the construction of the data set, since at any given F-scale rating, more tornadoes resulting in fewer casualties will be included in the more recent years. To address this problem, we could use the Grazulis record for more recent years, but this would involve discarding thousands of tornado records. Another possibility would be to choose an F-scale threshold for which we could be confident that all tornadoes meeting this threshold are included in the Grazulis record, which is what we did in our analysis of F5 tornadoes (Simmons and Sutter 2005a). One difficulty that arises from extending the data further back in time is that fewer variables are reported by the Census Bureau at the county level, meaning that the regression analysis can include fewer control variables. We would be less able to ensure that an observed time trend is not due to omitted variables. For these reasons, we chose not to extend our analysis earlier than 1950.

We performed the regression analysis to obtain a more conclusive answer to whether tornadoes have become less dangerous over time. The 1950–2007 analysis provides the best perspective on trends, as the 1986–2007 period constitutes only two decades. Figure 3.19 displays the time trend on casualties, calculated in each case using the year variable from the regressions. The figure reports the effect on casualties for a comparable tornado at the end of the period relative to the beginning of the period. Between 1950 and 2007, expected fatalities declined 42% while expected injuries declined 12%, with only the fatalities trend being statistically significant. Tornadoes have become less deadly over time, even when controlling for storm and storm-path characteristics, and the magnitude of the reduction is reasonably large. We cannot conclude with certainty that injuries have declined; if so, the magnitude of the decline in injuries is likely small. The differing results could be a consequence of more complete counting of injuries in recent years offsetting any decline in injuries. Over the shorter period 1986–2007, fatalities have

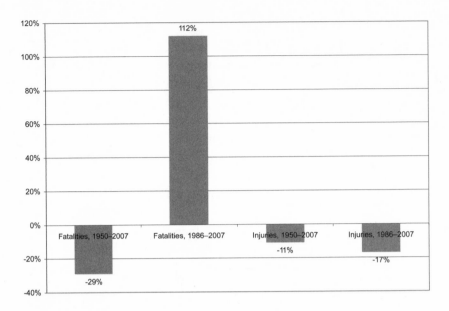

FIGURE 3.19. Regression results on time trends in casualties

actually increased, with the time trend indicating a more than doubling of fatalities for a comparable tornado. Injuries, by contrast, declined by a statistically significant 36%. Note that two decades represents a relatively short period in which to draw conclusions regarding time trends, however.

What does the regression analysis demonstrate regarding the effect of the various factors on casualties, holding other factors constant? We start with F-scale. Table 3.5 illustrates expected fatalities and injuries from tornadoes of various F-scale ratings relative to an F0 tornado, for the 1950–2007 and 1986–2007 regression models and the unadjusted differences in casualties per tornado. The table reports a rate for each F-scale category divided by the F0 rate; thus, the numbers can be interpreted as lethality relative to an F0 tornado. Fatalities increase more than injuries as F-scale increases, as indicated by the larger impact for each F-scale category. The two periods of regression analysis provide different results for fatalities relative to fatalities per tornado: Over 1950–2007, the regressions reduce the influence of F-scale, but over 1986–2007, F-scale has a greater influence on fatalities than in the univariate comparisons. The regression models over both time periods reduce the influence of F-scale on injuries, or alternatively the control variables explain some of the univariate influence of F-scale on injuries. The overall impact of F-scale remains large for both fatalities and injuries: Fatalities for an F5 tornado are four orders of magnitude greater than for an F0 tornado, and injuries are

TABLE 3.5. Expected Fatalities and Injuries by F-Scale Category

F-Scale	Fatalities			Injuries		
	Totals	1950–2007	1986–2007	Totals	1950–2007	1986–2007
1	11	11	32	11	9	11
2	57	43	200	49	37	64
3	473	298	1097	263	148	167
4	3402	1978	4273	1537	672	685
5	15141	8778	44356	4956	2165	1540

Totals refer to Fatalities/Tornado and Injuries/Tornado based on all tornadoes rated in the F-scale category from 1950 to 2007. 1950–2007 is based on the F-scale coefficients from the regression analysis of tornadoes over this period, and 1986–2007 is from the regression coefficients over this period. The regression models are presented in the appendix to this chapter. Rates are expressed relative to an F0 tornado.

three orders of magnitude greater for F5 than F0. The marginal impact of a one-category increase in F-scale on fatalities is relatively consistent across both regression models, with expected fatalities at least quadrupling in each case. The 1986–2007 analysis reveals a very large increase in lethality from F0 to F1 (perhaps because F0 tornadoes essentially no longer kill people), and an order of magnitude increase from F4 and F5. An order of magnitude increase in expected injuries is observed when moving from F0 to F1, and then generally a one-category increase in F-scale triples injuries. These results indicate that a rapid assessment of the strength of a tornado could greatly assist emergency managers and medical providers: For example, if the tornado that just struck a community is an F4 and not an F3, emergency managers can expect 3 to 4 times more fatalities and injuries.

Time of day matters substantially for casualties, with a similar pattern across the day for both fatalities and injuries, as illustrated in Figures 3.20 and 3.21. These figures are reported as an index we created to illustrate the impact of time of day. The index normalizes casualties for early afternoon tornadoes to 100. An index value of 150, then, indicates that expected casualties are 50% greater than for a comparable tornado in the early afternoon, while values less than 100 indicate lower expected casualties. Tornadoes after dark are significantly more likely to kill or injure people. The impact of time of day on both fatalities and injuries is more pronounced in the 1986–2007 analysis. Relative to an otherwise similar tornado in the early afternoon, fatalities are 48% and 42% higher overnight and during the late evening, respectively, based on the 1950–2007 analysis, and 153% and 90% higher based on the 1986–2007 analysis. Injuries depend less on time of day than

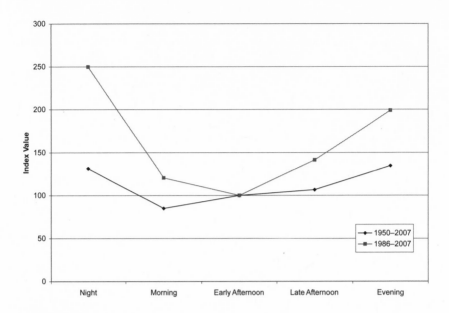

FIGURE 3.20. Regression analysis of fatalities and time of day

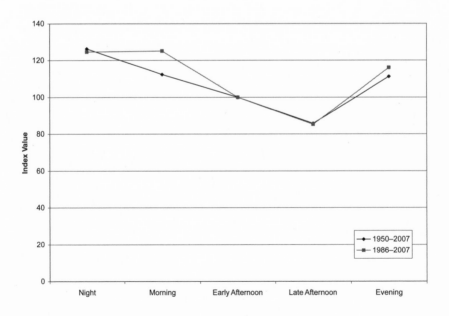

FIGURE 3.21. Regression analysis of injuries and time of day

fatalities, with expected injuries 26% and 15% higher overnight and in the late evening, respectively, in the 1950–2007 analysis, and 60% and 37% higher in the 1986–2007 analysis. Residents are probably less likely to hear a tornado warning or approaching tornado at night and thus less likely to take cover, which may account for the difference in vulnerability. The greater influence of time of day on fatalities than injuries suggests that taking cover is more likely to allow a person to avoid a fatal injury than avoid injury at all.

The differences in lethality of tornadoes across the year hold up in the regression analysis when controlling for other determinants of casualties. Figures 3.22 and 3.23 again display an index for casualties, with the value for July here set to 100. Index values above and below 100 thus indicate greater or lesser vulnerability relative to July. Tornadoes in the late spring and summer are less dangerous than at other times of the year, particularly the "off-season" or winter months. The regression analysis is particularly important for month effects because the casualties per tornado reported in Figures 3.6 and 3.7 did not account for differences in the F-scale distribution of tornadoes across the year. The influence of month on casualties is substantial, as expected fatalities in many months exceed July by a factor of three, and in the 1986–2007 analysis, February fatalities are more than seven times the July rate. Expected injuries also vary across the months by a factor of three. The regression analysis does not explain the source of this differential in vulnerability, but conceivably tornadoes in the "off-season" months are more likely to catch residents by surprise. Residents might expect that an ominous-looking spring thunderstorm could produce a tornado and be alert for a warning, but discount the tornado-producing potential of a fall thunderstorm.

Figure 3.24 displays the estimated impacts for weekends versus weekdays, and reveals that tornadoes on the weekend are more dangerous. Fatalities are 22% higher for weekend tornadoes based on the 1950–2007 analysis and 31% higher based on 1986–2007; injuries are only about 8% higher on weekends in each period, an increase that is barely statistically significant. The statistical analysis indicates that of the various factors discussed above regarding the lethality of weekday versus weekend tornadoes, factors such as the relative safety of businesses and schools (recall from Figure 3.14 that only about 5% of fatalities occur in businesses and schools, respectively) and residents being less likely to receive warnings on weekends predominate.

Population density increases injuries significantly in both samples. For fatalities, the direct impact of density was modest, but it did significantly increase fatalities through the interaction with path length. Figure 3.25 displays the impact of an increase in population density of 1,000 persons/mi^2

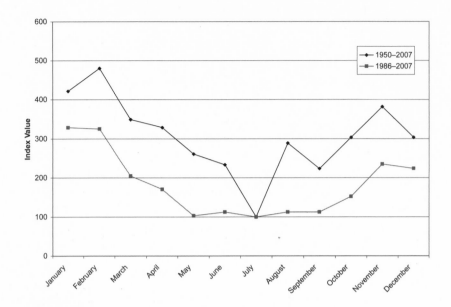

FIGURE 3.22. Regression results for fatalities and month

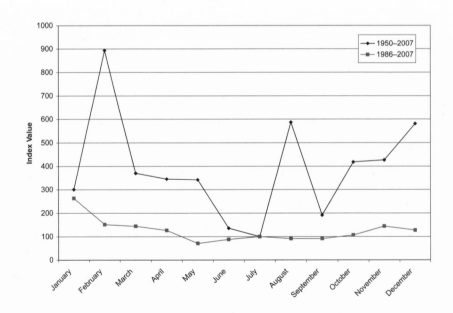

FIGURE 3.23. Regression results for injuries and month

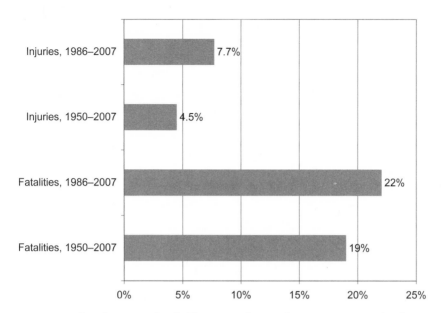

FIGURE 3.24. Casualties on weekends. The impact of a tornado occurring on a weekend, relative to a weekday, calculated from the regression results in the chapter appendix.

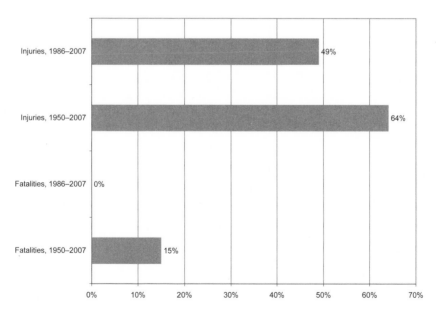

FIGURE 3.25. Population density and casualties. The figure reports the impact of a 1,000-persons-per-square-mile increase on expected casualties, calculated from the regression results in the chapter appendix.

on casualties; note that this is a very large difference in population, and essentially amounts to a shift from a rural area to an urban or suburban setting. A 1,000 persons/mi^2 increase in density increases expected fatalities by about 10% in each set of regressions, while injuries increase by 65% in the 1950–2007 sample and by 49% in the 1986–2007 model. A difference in population density of this magnitude is due to urban versus rural counties, not population growth; U.S. population density only increased from 43 persons/mi^2 to 83 persons/mi^2 between 1950 and 2000, while only those areas that transitioned from rural to suburban have experienced an increase in population density approaching 1,000 persons/mi^2. Population affects casualties, but its affect depends on whether the tornado strikes an urban area, not overall population growth.

Figure 3.26 displays the effect of a $10,000 increase in median family income on expected casualties in the two models. This increase in income reduces expected fatalities by 1% and increases expected injuries by 7% in the 1950–2007 model, and only the injuries effect is statistically significant. Over the 1986–2007 period, a $10,000 increase in income reduces fatalities by 6% and increases injuries by 44%, although again only the injuries effect is statistically significant. The 1986–2007 analysis includes several variables associated with income that significantly increase fatalities. Our previous research found that income increased fatalities in tornadoes since 1986 (Simmons and Sutter 2005, 2008), but this analysis did not include home prices or home age. We find no evidence that tornado safety is a luxury good, as other research on natural hazards has found. The increase in injuries we found may be because wealthier households are more likely to seek medical treatment for less severe injuries, or because emergency managers in wealthier communities exert more time and effort to report injuries.

Many of the other economic and demographic variables attain significance in the 1986–2007 analysis. Our discussion here will report the impact of a one standard deviation increase in the control variable on fatalities or injuries when statistically significant. A one standard deviation increase in the proportion of mobile homes (.08) increases expected fatalities by 32% and expected injuries by 22%. In many counties mobile homes comprise 25% or more of the housing stock, so mobile homes contribute substantially to casualties. The age of homes also affects casualties, as a one standard deviation increase in home age (11 years) reduces fatalities by 32% and injuries by 17%. Improvements in technology, construction techniques, and materials would plausibly reduce casualties, but at least in terms of tornadoes, newer is not necessarily better.

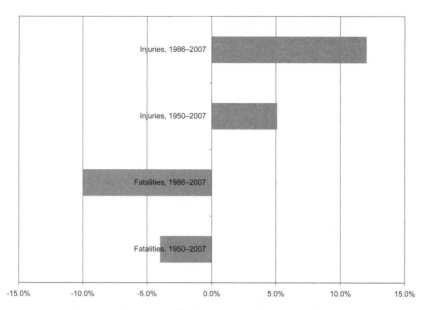

FIGURE 3.26. Income and casualties. The figure reports the impact of a $10,000 increase in median family income on expected casualties, calculated from the regression results in the chapter appendix.

A larger rural population reduces casualties, everything else held constant. A one standard deviation increase in rural population decreases expected fatalities by 22% and expected injuries by 11%. The greater concentration of population in urban areas more than offsets the greater potential for a tornado to strike a less populated area, holding population density constant. An increase in the proportion of workers with a commute in excess of 30 minutes does not significantly affect fatalities, but reduces injuries by 9%.

Increased educational attainment reduces casualties. A one standard deviation increase in college education reduces expected fatalities by 20% and expected injuries by 23%. The low level of educational attainment reflected by the lack of a high school degree did not significantly affect fatalities, but a one standard deviation increase in high school dropouts increased injuries by 22%. An increase in the proportion of men in the county population reduces injuries. Conventional wisdom holds that the young and elderly are vulnerable to natural hazards, but we find no evidence of this for tornadoes. Indeed, a one standard deviation increase in the proportion of residents over 65 reduces expected fatalities by 26% and injuries by 24%, while an increase in the population under 18 decreases injuries by 22%. The over-65 results run counter to the age breakdown of tornado victims: More than 25% of victims were aged 65 or older, compared to only 12% of the U.S. population, consistent with the vulnerability of the elderly. The regression result shows that an increase in the elderly population of a county, everything else held equal, lowers the expected number of deaths, even though for a given number of deaths, the victims tend to be older.

The vulnerability of these groups has received great attention since Hurricane Katrina, and Donner (2007) finds some evidence of such vulnerability for tornadoes. We find no support for the vulnerability of the poor or minorities to tornadoes: In our analysis, the poverty rate and proportion of nonwhite residents are not significant determinants of fatalities or injuries. In addition, the point estimates are small, with a one standard deviation increase in these variables never increasing or decreasing casualties by more than 7%. While specific examples of the vulnerability of poor or minority communities to tornadoes exist (Aguirre 1988), our analysis finds no systematic relationship between poverty or minority populations and tornado casualties.

3.5. A Closer Look at Tornado Vulnerability

Our analysis of tornado casualties has identified several sources of heightened lethality, including mobile homes, tornadoes at night, and tornadoes during the "off-season" or winter months. Our regression analysis demonstrates that these factors increase fatalities and injuries when everything else is held constant. But regression analysis does not allow us to see exactly *how* these factors increase casualties. Intuition suggests plausible sources of the mobile-home and nighttime vulnerabilities, but intuition can be misleading and does not substitute for close analysis. In this section we examine these high vulnerability circumstances for insights on casualties. The 1986–2007 period regressions controlled for more potential determinants of casualties and still found strong nocturnal and winter vulnerabilities, so we focus on tornadoes and casualties from this period.

The NWS records on tornado fatality locations from 1996 to 2007 allow us to explore the "mobile-home problem" in greater detail. We first consider the distribution of all fatalities and mobile-home and permanent-home fatalities by state. Table 3.6 reports the breakdown of fatalities by location for states with at least 10 fatalities over the 1996–2007 period, since a difference in the distribution of fatalities by location is not very meaningful when the total number of fatalities is too small. Nonetheless, a single tornado accounts for half or more of fatalities in several states in Table 3.6, so we do

TABLE 3.6. Fatalities by Location across States

State	Fatalities	Percentage in Mobile Homes	Percentage in Permanent Homes	Percentage in Other Locations
Alabama	88	44.3	43.2	12.5
Arkansas	56	46.4	28.6	25.0
Florida	75	70.7	10.7	18.7
Georgia	59	86.4	5.1	8.5
Illinois	20	35.0	5.0	60.0
Indiana	27	92.6	0.0	7.4
Kansas	29	20.7	58.6	20.7
Kentucky	11	72.7	18.2	9.1
Louisiana	18	72.2	5.6	22.2
Mississippi	17	23.5	58.8	17.6
Missouri	48	25.0	56.3	18.8
North Carolina	15	86.7	6.7	6.7
Ohio	11	27.3	36.4	36.4
Oklahoma	45	28.9	60.0	11.1
Tennessee	87	48.3	41.4	10.3
Texas	59	32.2	57.6	10.2
Total	729	48.1	35.0	16.9

Based on tornadoes during 1996–2007. Only states with at least 10 fatalities over the period are listed.

not want to infer too much from the location of fatalities in all these states. For example, 24 of the 27 fatalities in Indiana occurred in the November 2005 Evansville tornado, which struck a mobile-home park. Indiana had no permanent-home fatalities over this period, but we would not want to conclude that permanent-home residents are not in danger in the Hoosier state. Sixteen states had at least 10 fatalities over the years, including eight with 45 or more, so we have several states with fatalities from a number of killer storms. The distribution of fatalities by location differs significantly (at the .10 level) from the national distribution over these years for 11 states: Florida, Georgia, Indiana, Louisiana, and North Carolina, due to a large proportion of mobile-home fatalities; Kansas, Mississippi, Missouri, Oklahoma, and Texas, because of a higher proportion of permanent-home fatalities; and Illinois, due to a large proportion of fatalities in other locations. Both Florida and Georgia had more than 45 fatalities and a high proportion of mobile-home fatalities, and so these two states contributed greatly to the mobile-home problem. The states with a high incidence of permanent-home fatalities all had violent tornadoes during this period, including F5 tornadoes in Kansas, Oklahoma, and Texas. In addition, Alabama, Arkansas, and Tennessee, which had proportions of fatalities close to the nationwide average, experienced F4 or F5 tornadoes.

Mobile-home fatalities have a strong Southeastern component. Nearly 30% of all U.S. mobile-home fatalities occurred in Florida and Georgia, and 58% in the Southeastern states of Alabama, Florida, Georgia, Mississippi, North Carolina, and South Carolina. For perspective on the number of mobile-home fatalities in Florida and Georgia, we can calculate the number of mobile-home fatalities expected in these states based on fatalities that occurred in locations other than mobile homes nationally. Between 1996 and 2007, 51.9% of U.S. fatalities occurred in locations other than mobile homes. Florida had 22 fatalities in locations other than mobile homes; if these 22 fatalities totaled 51.9% of Florida fatalities, the state would have had 20 mobile-home fatalities. Instead, Florida experienced 53 mobile-home fatalities, or 33 more than expected. Georgia is even more extreme. Eight fatalities occurred in locations other than mobile homes in Georgia, so we would expect about 7 mobile-home fatalities in the state, not the 51 actually observed. The mobile-home problem clearly contributes to the overall Southeastern vulnerability to tornado hazards documented by Boruff et al. (2003) and Ashley (2007).

Figure 3.14 reports the proportion of fatalities in mobile and permanent homes by F-scale category over the period 1996–2007. If mobile homes and permanent homes are comparably vulnerable to tornadoes, the proportion of fatalities in mobile homes should be similar for weak, strong, and violent tornadoes. Nationally, 48% and 35% of fatalities occurred in mobile and permanent homes over the period. Only 4 fatalities occurred in F0 tornadoes, which is too small to draw any inferences, so when we ignore this category we clearly see that the proportion of fatalities in mobile homes decreases and the proportion of fatalities in permanent homes increases as we increase the F-scale ratings from F1 to F5. Mobile homes account for 70% of F1 fatalities, and 60% of F2 fatalities, but only 14% of F5 fatalities, while the percentage in permanent homes rises from 8% for F1, 23% for F2, 26% for F3, 56% for F4, to 83% for F5. The distribution of fatality locations differs significantly from the national distribution for all F-scale categories except F0 and F4. The disproportionate share of fatalities in mobile homes in F1, F2, and F3 tornadoes is consistent with the greater vulnerability of these homes to less powerful tornadoes. Permanent-home fatalities, by contrast, generally occurred in violent tornadoes. Permanent homes seemingly provide enough

TABLE 3.7. Tornadoes and Killer Tornadoes by Time of Day, 1996–2007

	Proportion of All Tornadoes	Proportion of All Killer Tornadoes	Proportion of All Mobile Home Killer Tornadoes
Midnight–5:59 AM	.0611	.1402	.2138
6 AM–11:59 AM	.0830	.0379	.0621
Noon–3:59 PM	.2674	.2083	.2000
4 PM–7:59 PM	.4590	.3750	.3034
8 PM–11:59 PM	.1294	.2386	.2207

The distribution of times of killer tornadoes and mobile-home fatality tornadoes differ from the distribution of for all times in a chi-square test (p-values < .001).

protection to avoid fatal injuries in F3 and weaker tornadoes, while residents of mobile homes face mortal peril in these tornadoes.

We next break down mobile-home fatalities based on the timing of tornadoes, using the five categories for the parts of the day employed earlier: overnight (midnight to 5:59 AM), morning (6:00 AM to 11:59 AM), early afternoon (noon to 3:59 PM), late afternoon (4:00 PM to 7:59 PM), and evening (8:00 PM to 11:59 PM, all times local). Table 3.7 displays the distribution of all tornadoes, killer tornadoes, and mobile-home fatality tornadoes by parts of the day. Tornadoes are most frequent in the late afternoon (46%), followed by the early afternoon (27%), evening (13%), morning (8%), and overnight (6%). Killer tornadoes are much more likely after dark; the proportion of killer tornadoes during the evening and overnight hours (38%) is double the proportion of all tornadoes during these times (19%), and tornadoes producing mobile-home fatalities are even more likely after dark (43%). Fatality and mobile-home killer tornadoes are both less likely during the early and late afternoon than tornadoes in general. The distributions of killer tornadoes and mobile-home fatality tornadoes differ statistically from the distribution of all tornadoes.[8]

Nocturnal and mobile-home fatalities appear closely intertwined. Figure 3.27 displays the distribution of fatalities (as opposed to tornadoes producing fatalities) in mobile and permanent homes by time of day. One-third of mobile-home fatalities occur during the overnight hours, versus only 5% of permanent-home fatalities. More than 3 out of 4 overnight fatalities occur in mobile homes. By contrast, 54% (25%) of permanent- (mobile-) home fatalities occur during the late afternoon, and half of all late afternoon fatalities occur in permanent homes. Tornadoes occur more frequently during the late afternoon, and violent tornadoes are particularly likely in the late afternoon or evening. Also of note, equal proportions of fatalities occur in

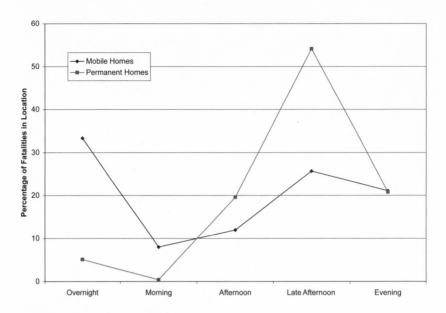

FIGURE 3.27. Mobile-home and permanent-home fatalities by time of day. Based on tornadoes 1996–2007. The percentages show the distribution of fatalities in each location across the day-parts defined in the text.

each location in the late evening, while more mobile- (permanent-) home fatalities occur during the morning (afternoon).

Why do nighttime tornadoes kill so many residents of mobile homes? Figure 3.28 displays mobile- and permanent-home fatalities per tornado at the different times of day to try to further explore this question. About one mobile-home fatality occurs per eight tornadoes during the overnight hours, a rate that is about three times greater than the next highest rate for either location at any other time of the day. Permanent-home fatalities per tornado are slightly lower overnight (.014) than for all tornadoes (.017). Nocturnal tornadoes appear deadlier only for mobile homes. The warning dissemination and response process likely does not perform as well at night, as residents might fail to receive a warning because they are asleep, or be less likely to respond to a warning if they cannot visually confirm the danger.[9] If the warning process explained nocturnal vulnerability, we would expect to see fatalities per tornado increase for both mobile and permanent homes. But as Figure 3.28 shows, the fatality rate increases at night only for mobile homes. A more plausible explanation might be the vulnerability of mobile homes combined with residents more likely to be at home (instead of at

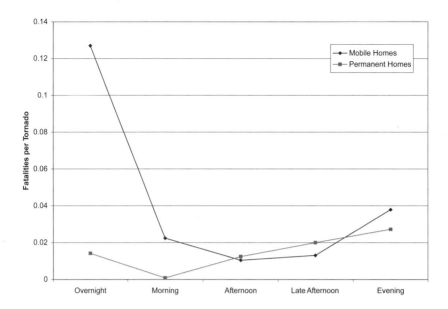

FIGURE 3.28. Fatalities per tornado by location and time of day. Based on tornadoes 1996–2007. The numbers are total fatalities in each location divided by the number of tornadoes in each day-part.

work or school) during the overnight hours. Fatalities per tornado is not conclusive on this score, as it does not account for factors like the F-scale distribution of nighttime tornadoes or the characteristics of tornado paths, but it is certainly suggestive.

The nocturnal dimension of the mobile-home problem is consistent with a recent study by Schmidlin et al. (2009) of the warning response of mobile-home residents. They found that 69% of mobile-home residents did not take shelter upon receiving a warning, and further that almost one-third of responders failed to choose an option that reduced their vulnerability. Schmidlin et al. focused on warnings issued between 7 AM and 11 PM to ensure that they could find respondents who were home and received the warnings, but the lack of response to warnings likely applies to nighttime tornadoes as well. If mobile-home residents are likely to remain in their homes and are more likely to be home for overnight tornadoes, this could explain the nighttime mobile-home problem.

We next turn to the greater lethality of tornadoes during the overnight period, midnight to 6 AM. A total of 1,398 tornadoes occurred during these hours from 1986 to 2007, in 40 different states, with 241 of the tornadoes rated F2. The 5.6% of tornadoes occurring overnight accounted for 172 fatalities and 2,648 injuries, or 15% of all fatalities and 12% of all injuries.

Why are these tornadoes so lethal? Figure 3.29 reports the ratio of fatalities and injuries per tornado for the overnight period to other times of the day, for each F-scale category (except F5, as no F5 tornadoes occurred during the overnight hours in these years). A ratio greater than 1 indicates that casualties are greater at night than at other times. As we might expect, fatalities and injuries per tornado are greater at night for each F-scale category, but nighttime tornadoes result in relatively more casualties for weaker tornadoes. At the extreme are F0 tornadoes, for which the fatality rate is almost an order of magnitude greater at night than at other times. Of course, F0 tornadoes produce few fatalities (8 in 22 years), so the base casualty rates are low and thus the relative lethality of F0 tornadoes accounts for very few deaths. Figure 3.29 also shows that F3 tornadoes are more dangerous at night relative to F1, F2, or even F4 tornadoes: F3 tornadoes are about three times more dangerous at night, while F2 and F4 tornadoes are about twice as dangerous. F3 tornadoes accounted for 60% (103) of nighttime fatalities and just over half of nighttime injuries, so much of the nocturnal vulnerability is due to F3 tornadoes.

We also examined nighttime fatalities by state. Seventeen states experienced at least one nighttime tornado fatality over the 1986–2007 period, with the most occurring in Georgia (31), followed by Florida (29), Indiana (24, all in the November 2005 Evansville tornado), Alabama (19), North Carolina (17), and Mississippi (14). Figure 3.30 displays the percentage of tornadoes, fatalities, and injuries at night for the 10 states that experienced 5 or more nighttime fatalities over the period; these states account for 92% of nighttime fatalities. A state exhibits nocturnal vulnerability when the percentage of nighttime casualties exceeds the percentage of nighttime tornadoes. For instance, Alabama has the fourth highest nighttime fatality total, but about 13% of the state's tornadoes, fatalities, and injuries occur during these hours, so overall nighttime tornadoes do not appear to be more lethal in Alabama than tornadoes at other times of the day. As evaluated in this manner, nighttime tornadoes are most lethal in Indiana and North Carolina, where about 5% of tornadoes occurred at night but accounted for over 55% of fatalities. In Indiana, the November 2005 Evansville tornado drives nighttime fatalities, so whether Indiana exhibits greater night vulnerability or just experienced one particularly deadly nighttime tornado is unclear. Georgia and Mississippi also evidence considerable nighttime vulnerability, as the 13% of Georgia tornadoes that occur at night result in about one-third of the

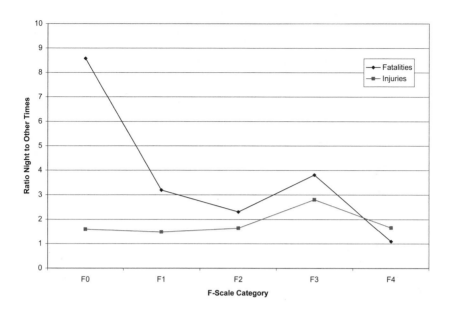

FIGURE 3.29. Relative lethality of nighttime tornadoes by F-scale

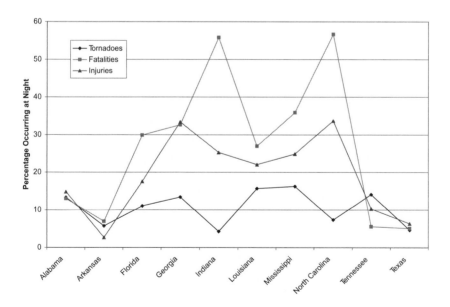

FIGURE 3.30. Nighttime tornadoes and casualties by state. The percentages are of all tornadoes, fatalities, or injuries in the state occurring between midnight and 6 AM. States displayed had at least 5 nighttime fatalities.

TABLE 3.8. The Regional Component of Nighttime Fatalities

	Southeastern States	All Other States
Fatalities per Tornado	.187	.071
Injuries per Tornado	2.85	1.12
Fatalities per Strong/Violent Tornado	.806	.420
Injuries per Strong/Violent Tornado	12.1	6.61

Totals are for tornadoes between midnight and 6 AM, local time. The Southeastern states included here are Alabama, Florida, Georgia, Mississippi, North Carolina, South Carolina, and Tennessee.

state's fatalities and injuries, and the 16% of Mississippi tornadoes that occur at night account for 36% of state fatalities. In Florida, the 11% of tornadoes occurring at night account for 30% of fatalities but only 17% of injuries; nocturnal tornadoes are quite lethal in the Sunshine State, yet only modestly increase the injury total.

Examination of the states in Figure 3.30 identifies a Southeastern component to nighttime fatalities, with four high vulnerability states: Florida, Georgia, Mississippi, and North Carolina. To test the regional component of nighttime vulnerability, Table 3.8 reports fatalities and injuries per tornado for all F2+ nighttime tornadoes for the seven Southeastern states (these four plus Alabama, South Carolina, and Tennessee) and then for all other states. Nighttime tornadoes definitely appear more dangerous in the Southeast, with casualties per tornado are approximately double the national rate. The difference in casualty rates is slightly greater for weak tornadoes than for F2 and stronger tornadoes, but the low base casualty rate for weak tornadoes implies that this does not explain many of the Southeastern casualties. Note that since Figure 3.30 displays rates per tornado, the greater frequency of tornadoes at night in the Southeast does not drive the difference in vulnerability. The nocturnal tornado problem, like the mobile-home problem, has a strong Southeastern regional component.

We now turn to the variation in casualties across the year: the "off-season" or winter tornado vulnerability. The off-season vulnerability strikes us as much less intuitive than the mobile-home or nighttime vulnerability. Although tornadoes during the off-season might catch residents by surprise, the variation in casualties between July and February exceeds the variation observed for night versus daytime tornadoes. The winter vulnerability thus provides a perplexing puzzle. To explore winter vulnerability, we focus on tornadoes during the months of November, December, January, and Februar-

TABLE 3.9. Casualty Rates by State, for November–February Tornadoes

State	All Tornadoes		F2+ Tornadoes	
	Fatalities/Tornado	Injuries/Tornado	Fatalities/Tornado	Injuries/Tornado
Alabama	.234	4.93	.951	17.8
Arkansas	.122	1.90	.437	6.44
Florida	.260	2.21	2.52	18.2
Georgia	.188	3.55	.778	13.5
Indiana	.667	7.67	1.50	17.0
Kentucky	.091	2.55	.118	6.52
Louisiana	.028	1.39	.091	6.04
Mississippi	.084	2.53	.298	10.1
Missouri	.070	1.73	.400	10.3
New York	.900	1.90	N.A.	N.A.
North Carolina	.229	6.09	1.08	27.2
Ohio	.135	1.62	.357	4.07
Tennessee	.320	3.76	.673	8.87
Texas	.040	1.05	.278	6.50
All States with Fatalities	.137	2.45	.590	10.2

ary. About 12% of tornadoes nationally over the 1986–2007 period occurred in these months, with November accounting for the most winter tornadoes, at 45% of the total during those four months. Winter tornadoes occurred in 38 states, although 12 states had fewer than 10. Winter tornadoes accounted for 32% of all fatalities and 31% of injuries over the period, and fatalities occurred in 17 states, led by Florida (69), Alabama (60), and Tennessee (40).

As suggested by the states with the largest fatality totals, the off-season vulnerability also has a strong Southeastern flavor. Table 3.9 reports fatalities and injuries per tornado for states with five or more off-season fatalities. Fatalities per tornado in the 17 states with winter fatalities were .137 overall, and .590 per F2 or stronger tornado. Fatalities per tornado vary widely across states, with the highest rates in New York and Indiana. New York's vulnerability resulted from 9 fatalities in 10 weak winter tornadoes, while Indiana's 24 winter fatalities all occurred in the November 2005 Evansville tornado. The next highest fatality rates for winter tornadoes were in Tennessee, Florida, Alabama, and North Carolina, again demonstrating the Southeastern component of off-season vulnerability. North Carolina and Indiana have the highest injury rates, while Florida and New York actually have injury rates

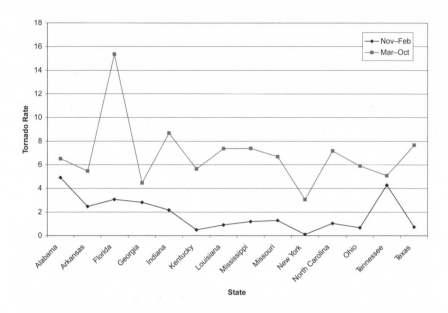

FIGURE 3.31. Tornado rate by state and month

below the average for all states with winter tornado fatalities. Thus different factors seem to affect fatalities and injuries in the winter. Figures 3.31 and 3.32 further examine winter tornadoes in these states. States might experience a large number of winter fatalities either because tornadoes occur more frequently in these months or because tornadoes are more dangerous in these months. Figure 3.31 reports the annualized tornado rate per 10,000 mi^2 for the winter months and for March through October for states with five or more winter fatalities. The tornado rate for the rest of the year exceeds the winter rate in every state, and only in Tennessee are the rates almost equal. Florida experienced the most winter fatalities over the period, but the state's winter tornado rate was about one-fifth of the rate during the rest of the year. The 18% of Florida tornadoes occurring between November and February account for two-thirds of the state's tornado fatalities. Of course, not all tornadoes are equally powerful. To see if a difference in powerful tornadoes affects casualties, Figure 3.32 reports the percentage of tornadoes rated F2 or stronger during the winter months and the other months in each state. A larger percentage of tornadoes are rated F2 or stronger in the winter months than during the rest of the year in each state except New York. Over 40% of winter tornadoes in Indiana and Tennessee are strong or violent, while in no state were more than 25% of tornadoes in the rest of the year rated F2 or stronger. Only about 10% of Florida winter tornadoes were rated F2+,

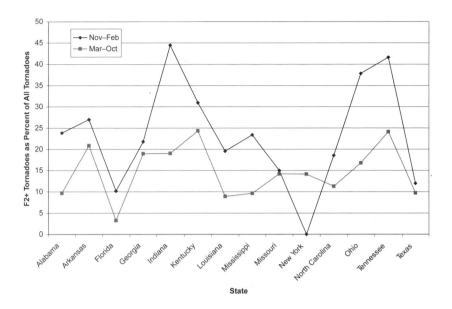

FIGURE 3.32. Percentage of strong and violent tornadoes by state and month

although this exceeded the percentage for the rest of the year. Note, however, that more F2 and stronger tornadoes occurred between March and October (40) than in the winter months (27). Some of Florida's winter tornado vulnerability might be due to part-year winter residents.

To further control for the strength distribution across the year, Figure 3.33 displays the ratios of fatalities and injuries per tornado during the winter to other months, by F-scale category. (The F5 category is omitted because no F5 tornadoes occurred during the winter over these years.) The ratio exceeds 1 for each F-scale category, indicating that winter tornadoes result in more casualties; thus, the off-season vulnerability is pervasive. The highest ratios for fatalities and injuries in Figure 3.33 occur in the F0 and F1 categories, so weak tornadoes are more dangerous compared with strong or violent tornadoes during the winter. Again, the casualty rates are absolutely low for weak tornadoes, so the greater lethality translates into few extra casualties. Winter F4 tornadoes are only about 35% more lethal, and so the off-season vulnerability is definitely more pronounced for weaker tornadoes. Winter F3 tornadoes are considerably more dangerous, and due to the higher rate of fatalities, 55% of fatalities in November through February have occurred in F3 tornadoes.

Shorter days offer one possible explanation for the lethality of winter tornadoes (Ashley et al. 2008). Tornadoes after dark are more dangerous,

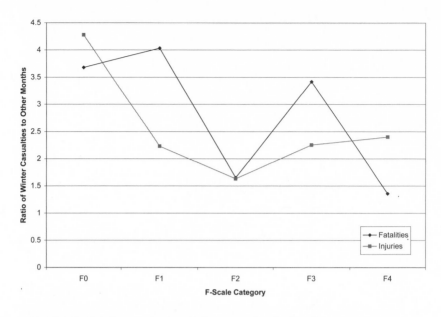

FIGURE 3.33. Relative casualty rate of winter tornadoes, by F-scale category

everything else being equal, and due to shorter days, a tornado at 7 PM is after dark in December but before sunset in June. Ashley et al. (2008) controlled for the effect of shorter days by comparing the timing of each tornado to the local time of sunset. We calculate casualty rates by parts of the day for tornadoes during the winter and other months of the year to explore this possibility. Figure 3.34 displays the ratio of fatalities and injuries per tornado for winter months to fatalities and injuries per tornado in other months. A ratio of greater than 1 indicates that winter tornadoes are more dangerous. We are particularly interested in the 4–6 PM, 6–8 PM, and 8–10 PM periods, since these are the hours when earlier sunsets might contribute to greater lethality for winter tornadoes. Figure 3.34 demonstrates that winter tornadoes are more dangerous throughout the day, with fatality and injury rates at least double the rates during the rest of the year for all parts of the day. The smallest increases in the lethality of winter tornadoes occur during the overnight, morning, and early afternoon time periods; this is expected, since the change in the time of sunset does not affect these tornadoes. Yet the winter fatality and injury rates are most elevated in the 10 PM and midnight periods, when they are 8 and 5 times higher than during the rest of the year, respectively. These tornadoes occur after dark throughout the year, so earlier sunsets do not appear to explain why tornadoes at this time are so dangerous in the winter. The greater lethality of off-season tornadoes across the day suggests

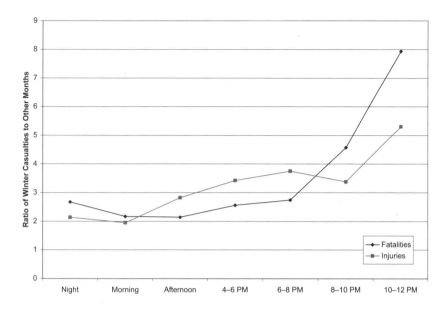

FIGURE 3.34. Relative casualty rate of winter tornadoes, by time of day

that the source of this vulnerability must extend beyond the difference in the timing of sunset.

Winter tornadoes are more lethal across F-scale categories in addition to time of day. And although the off-season tornado problem is worse in the Southeast, we observe elevated casualty rates across the country. One explanation for the pervasive elevated lethality of off-season tornadoes is a lulling effect, with residents failing to recognize the potential for tornadoes in the winter and being caught off guard. The frequency of tornadoes during the off-season (Figure 3.31) may fall below a threshold at which residents ignore the risk; they may not think that an ominous-looking thunderstorm in November or February can spawn a tornado. The diminished attention may lead residents to not be alert for, or even to dismiss, a tornado warning. Reduced attentiveness can potentially substantially increase risk for short-fuse warning events like tornadoes.

3.6. Are Tornadoes More Deadly in Some States?

Both the frequency of tornadoes and the incidence of fatalities and injuries vary greatly across states. Not surprisingly, many of the states with the highest rates of fatalities and injuries per million residents also have high rates of tornadoes and strong and violent tornadoes. For instance, the top five states in fatalities per million residents over the 1950–2007 period, Mississippi, Arkansas, Alabama, Kansas, and Oklahoma, all rank among the top eight states in the annual rates of and probability of damage from F2 or stronger tornadoes. Our analysis of tornado vulnerabilities in Section 3.5 found a strong Southeastern component to nighttime, mobile-home, and winter fatalities, which reinforces the potential for differences in casualties across states. A close comparison of Tables 2.7 and 3.1, however, reveals some differences between state ranks for tornado frequency and casualties. Michigan has a higher fatality rate than Iowa, even though Iowa ranks in the top five in each measure of tornado frequency while Michigan ranks between 15th and 30th. Illinois, Florida, and Massachusetts have similar fatality rates but very different tornado frequencies: Illinois ranks between 5th and 11th in frequency and Massachusetts between 20th and 23rd, while Florida has the highest tornado rate but much lower F2+ tornado rates and tornado probabilities. In other words, an otherwise similar tornado appears to result in more casualties in some states than others. Fatalities per tornado range from 0.68 in Massachusetts and 0.32 in Tennessee to 0 in the seven states without a tornado fatality. Fatalities by state in F2 and stronger tornadoes were tabulated but not reported. These casualties exhibit even greater variation, ranging from 2.4 in Massachusetts and 0.9 in Michigan and Tennessee, to 0 in six states. Tornado Alley states have large numbers of strong and violent tornadoes, but the powerful tornadoes in these states do not appear as deadly, with fatalities per F2+ tornado ranging from 0.38 in Kansas (18th among all states) to 0.37 in Texas, 0.31 in Oklahoma, and 0.15 in Nebraska.

Comparisons of casualties per tornado or per strong or violent tornado do not allow us to control for other determinants like the time of day, month of the year, or tornado-path characteristics, and thus could be misleading. The regression models of tornado casualties allow us to control for all of these other factors, and we can test for state effects on casualties by creating a set of state dummy variables that equal 1 for the state in which a given tornado struck and 0 otherwise. By including these state dummy (or state fixed effects) variables in casualties regressions, we test for an impact of the state on casualties when controlling for all of the other variables included in the regressions. A technical problem arises in states with zero or a very small number of casualties, and thus we combine states with fewer than 10 fatalities between 1950 and 2007 into one category of Other States, and use this as our omitted category in the regression analysis. Proceeding in this manner, we can create dummy variables for 28 states, which together account

TABLE 3.10. State Effects on Casualties

State	Fatalities 1986–2007	Fatalities 1950–2007	Injuries 1986–2007	Injuries 1950–2007	Casualty Effect Index
Alabama	10.65	6.26	3.32	5.13	1.63 [23]
Arkansas	5.48	8.90	1.70	3.96	1.01 [17]
Florida	27.54	12.71	3.16	5.06	2.56 [25]
Georgia	12.10	12.56	5.02	8.02	2.65 [26]
Illinois	5.19	5.13	2.46	2.98	0.90 [12]
Indiana	6.01	5.93	3.35	3.23	1.19 [20]
Iowa	1.13	1.05	1.79	1.96	0.30 [4]
Kansas	2.06	2.74	1.43	1.29	0.24 [3]
Kentucky	1.94	4.75	2.71	5.81	1.15 [19]
Louisiana	3.75	5.74	1.73	4.87	0.94 [16]
Massachusetts	9.55	36.69	5.56	4.10	3.08 [28]
Michigan	4.37	7.72	2.20	3.08	0.91 [14]
Minnesota	3.77	3.02	1.52	3.18	0.58 [7]
Mississippi	2.80	5.60	2.20	4.25	0.91 [13]
Missouri	4.86	3.78	2.48	3.06	0.85 [10]
Nebraska	1.06	1.20	1.45	1.91	0.22 [2]
New York	20.10	5.66	5.26	7.46	2.72 [27]
North Carolina	5.79	6.76	2.65	5.55	1.34 [21]
North Dakota	1.00	1.47	1.00	1.00	0.02 [1]
Ohio	3.31	5.17	1.83	4.94	0.93 [15]
Oklahoma	3.64	2.91	1.87	2.62	0.57 [6]
Pennsylvania	8.84	6.52	1.79	3.46	1.02 [18]
South Carolina	4.41	5.70	2.55	6.58	1.35 [22]
South Dakota	2.91	1.00	1.54	1.88	0.31 [5]
Tennessee	7.54	6.68	2.38	8.89	1.78 [24]
Texas	5.98	5.43	1.69	2.46	0.71 [8]
Virginia	8.83	2.75	1.49	2.71	0.72 [9]
Wisconsin	4.97	3.22	2.54	3.18	0.86 [11]
Other States	2.64	2.47	1.25	2.76	0.25
Average	6.41	6.32	2.45	4.02	1.09

for 90% (95%) of all tornadoes (tornadoes rated F2 or stronger).[10] We use this same set of state dummy variables for both fatalities and injuries, and for both the 1950–2007 and 1986–2007 periods of analysis. We estimate state effects in four regression specifications, Poisson models of fatalities over 1950–2007 and 1986–2007, and Negative Binomial models of injuries over these two periods. We can then compare the estimated state effects for both fatalities and injuries, and over both the longer period and since 1986, when we include many more control variables. The extra control variables in the 1986–2007 data set reduce the number of other factors that might correlate with state effects. A consistency of state effects across the different data sets and for both fatalities and injuries would be more persuasive evidence of some unchanging state-specific factor affecting tornado casualties.

The state dummy variables perform well in all four specifications, with many attaining statistical significance. Instead of reporting the full specifications or the raw regression coefficients, we report in Table 3.10 a measure of relative casualties for each specification. Specifically, we calculate the effect on casualties for each state, and then divide the effect by the smallest of the state effects for that specification.[11] The numbers in Table 3.10 are then the ratios of expected casualties from a tornado in a state to expected casualties from a comparable tornado in the state with the lowest casualty effect. Thus, a value of 4 for a state in a given specification indicates that expected casualties are four times greater than in the lowest-risk state.

Table 3.10 demonstrates that state effects have a sizable impact on casualties. The table reports each individual state effect, the mean of these state effects (which allows us to refer to states as having greater or lesser than average vulnerability), and the effect for the Other States. State effects are greater for fatalities than for injuries, and both the fatalities and injuries effects are comparable in size between the two time periods. North Dakota has the smallest state effect on casualties in three out of the four cases, while South Dakota has the smallest effect in the other case (fatalities over 1950–2007). Expected fatalities vary by an order of magnitude across states relative to the Dakotas. For fatalities over 1986–2007, the average of the state fixed effects is 6.4, or expected fatalities more than six times the level in North Dakota. Florida has the greatest effect, with expected fatalities almost 28 times greater than North Dakota. New York surprisingly ranks second at 20 times the expected fatalities in North Dakota, with Georgia and Alabama exceeding 10 and Massachusetts just below 10. Iowa, Kansas, Kentucky, and Nebraska have expected fatalities of less than double those of North Dakota. For fatalities over the 1950–2007 period, Massachusetts has the greatest effect with a factor of almost 37, due largely to the inclusion of the 1953 Worcester tornado. Florida and Georgia are next, at over 12 times the effect of South Dakota. Iowa, Nebraska, and North Dakota have expected fatalities of less than 50% greater than South Dakota. Mississippi, which has the highest fatality rate per million residents since 1950, has expected fatalities about 5.5 times greater than South Dakota, which is less than the average of state effects. Tornadoes do not appear to be particularly deadly in Mississippi; instead, the state's fatality rate is due to exposure to numerous dangerous tornadoes. The state effects for injuries are much smaller in magnitude. Massachusetts has the highest state effect over 1986–2007, at 5.5 times the expected injuries in North Dakota, while Tennessee has the greatest injuries effect for 1950–2007 at almost 9 times the expected injuries in North Dakota. Georgia and New

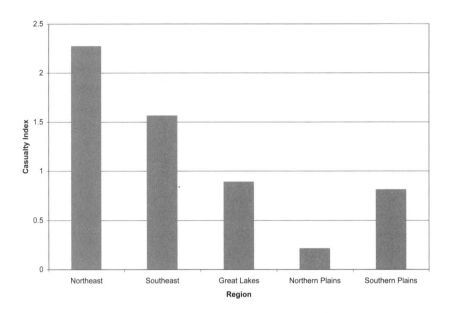

FIGURE 3.35. The lethality of tornadoes by region

York have large fixed effects on injuries over both 1986–2007 and 1950–2007. Kansas, Minnesota, Nebraska, South Dakota, and Virginia have the smallest effects on injuries over 1986–2007, while Iowa, Kansas, Nebraska, and South Dakota have the smallest impact over 1950–2007 (besides North Dakota).

The final column of Table 3.10 reports an index of the state's fixed effect on all four casualty regressions; a smaller index number indicates a smaller average fixed effect.[12] The states with the largest casualty index, Massachusetts and New York, come as no surprise based on the above discussion. The 1953 Worcester tornado contributes to the ranking for Massachusetts, but it is hardly the only factor, as Massachusetts also ranks high for fatalities over the 1986–2007 period. New York holds the second position despite modest totals of 21 fatalities and 307 injuries since 1950. The regression analysis indicates that tornadoes in New York result in more casualties than similar tornadoes in other states. After Massachusetts and New York, the Southeastern concentration of tornado casualties emerges as the next most vulnerable states are (in order) Georgia, Florida, Tennessee, Alabama, South Carolina, and North Carolina. The northern Plains states of (in order) North Dakota, Nebraska, Kansas, Iowa, and South Dakota have the smallest casualties index, with Oklahoma and Texas ranking 6th and 7th.

To confirm the regional patterns in state casualties, Figure 3.35 reports the average of the index for five groups of states. We divided the 28 states into groups we label Southeast, Northeast, Midwest, northern Plains, and southern Plains, with states assigned based on geography, and border states assigned according to their casualty index. The Northeast (Massachusetts, New York, and Pennsylvania) has the highest average of any region at 2.3, followed by the Southeast at 1.6, the southern Plains at 0.9, the Midwest at 0.8, and the northern Plains at 0.2. The vulnerability of the Northeastern states is difficult to explain, but tornadoes are sufficiently infrequent in these states that overall casualties are low. Also, the other New England states have sufficiently few (and in some cases no) fatalities that they are included in the Other States category, and so the average across all Northeastern states would be lower than reported in Figure 3.35. Three states on the border of the Southeast, Kentucky, Mississippi, and Virginia, have lower casualty indices and actually deflate the regional average in Figure 3.35. With these states removed, the Southeast average is 1.9, and no Southeastern states have sufficiently few tornado fatalities to fall into the Other States category. The vulnerability of the Southeast is significant because tornadoes are frequent enough that the Southeast accounts for a large proportion of the nation's tornado fatalities. The regional pattern can be further emphasized by graph-ing state fatality rates on a map; in Figure 3.36, we see a strong gradient of vulnerability from southeast to northeast. What do the state effects measure? The state dummy variables should capture factors that are state specific, unchanging over time, and not captured by our other control variables. The regressions control for the F-scale rating of the tornado, the time of day, and month of year differences in tornado climatology across states. Although the 1950–2007 regressions include relatively few control variables, the state ef-fects remain significant and consistent in the 1986–2007 regressions includ-ing a much richer set of storm-path characteristics. Thus, the state effects go beyond traditional demographic effects. Massachusetts's rank at the top of the list and influence of the 1953 Worcester tornado might suggest that the state variables are capturing older, more deadly tornadoes. However, the regressions include a time trend, and Massachusetts's state effect is also large in the 1986–2007 sample, and Michigan, which experienced the deadly 1953 Flint F5 tornado, does not have a particularly large state effect.

We conclude this section by examining the possible role of three tornado climatology features—annual tornado rate, coefficient of variation for the annual state tornado count, and the maximum monthly rate divided by the state annual rate—and two state features. The first factor is the frequency of the tornadoes. Learning how to prepare for a tornado is an investment: Residents must take the time to learn the difference between a watch and

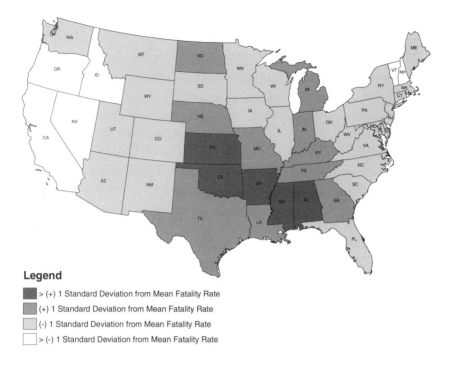

Legend

■ > (+) 1 Standard Deviation from Mean Fatality Rate
■ (+) 1 Standard Deviation from Mean Fatality Rate
□ (-) 1 Standard Deviation from Mean Fatality Rate
□ > (-) 1 Standard Deviation from Mean Fatality Rate

FIGURE 3.36. Tornado fatality rate by state. Map courtesy of Jeff Czajkowski.

a warning and where to take cover if a tornado approaches. Residents of more tornado-prone states might be more likely to prepare themselves, to buy NOAA Weather Radios, and so forth, reducing casualties from a given tornado. The annual tornado rate for all tornadoes (from Table 2.7) is negatively correlated with the state casualty index at −.12. Thus, tornadoes are less deadly in states with higher tornado rates, consistent with a threshold for tornado frequency for tornado preparations, but the relationship is modest and does not explain much of the state variation. The variance in tornado activity from one year to the next might also affect casualties. Consider two states that experience the same number of tornadoes over a five-year period, but the first state typically has one super tornado outbreak followed by several years of minimal activity, while the second state has a consistent number of tornadoes from year to year. After several years with few tornadoes, residents in the first state might be lulled into a sense of security and fail to respond quickly to the next outbreak. The coefficient of variation for the annual tornado count in the state (the standard deviation divided by the mean) is positively correlated with the state casualty index at +.32, so more variation in tornado threat across the year may affect casualties.

Conceivably, tornado awareness is very high during the season in Tornado Alley, leading to more effective response by residents. When tornadoes occur at a modest but steady rate throughout the year, the diffuse threat may result in a lack of focus in response by residents. In addition, the lack of a pronounced tornado season may affect residents' perceptions of their state's tornado rate. We saw in Chapter 2 that the traditional Tornado Alley states have a very concentrated, defined tornado season in the spring and early summer, while Southeastern states have a more constant rate across the year. We use the maximum annual equivalent monthly tornado rate divided by the state annual tornado rate reported in Table 2.8 to measure within-year concentration of the tornado threat. The correlation of the maximum rate with the state casualty index is −.39, so a concentrated tornado season, as in the Plains states, is less deadly.

We also consider the relationship between two state attributes—the average county size and forest cover in the state—and the casualty index. Until 2007, tornado warnings were issued for counties, and if the media alert residents based on counties, the size of the county will affect the quality of the alarm (we return to this point in Chapter 4). The correlation between average county size in a state (state land area divided by the number of counties) and the casualty index is −.38, indicating that states with larger counties have a smaller impact on casualties. While this runs counter to the effect of county size on the degree of risk conveyed, it seems to result from the low casualty index values for the Plains states, which also have large counties. Forest cover can affect the ability of residents to see an approaching tornado (Ashley 2007), thus reducing response to tornado warnings. The percentage of a state's land area covered by forest as of 2002 is highly correlated with the casualty index at +.62. The northern Plains states with the lowest casualty index values also have the least forest cover in the United States: only 1% of North Dakota land area is forest, and Kansas, Nebraska, and South Dakota each have less than 5% forest. Alabama and Georgia are the top two states with estimated casualty index values in forest cover. Forest cover is also statistically significant in a regression analysis of the state casualty index, explaining about 32% of the variation in the casualty index. None of the other four measures explored here attains significance in regression analysis once forest area is included as a control variable. Our results suggest that an inability of residents to see an approaching tornado increases the lethality considerably; this is also a possible link to the vulnerability to nocturnal tornadoes discussed in Section 3.5. Note that we have applied only a cross-sectional snapshot of forest cover and at the state and not the county level, and that forest land might have varied over time. Thus, this finding should be interpreted more as a suggestion for future research rather than a definitive finding.

3.7. The Lethality of Tornadoes: Fatalities by Location

On May 4, 2007, an F5 tornado devastated the town of Greensburg, Kansas. The tornado was the first to be rated EF5 since the adoption of the Enhanced Fujita Scale, and the first F5 tornado in the United States in eight years. The tornado leveled the town's business district and damaged or destroyed over 1,400 buildings. Figure 3.35 shows the devastation from this tornado. First responders to the tornado initially feared that hundreds of persons had been killed in the tornado, because based on the devastation, it seemed unlikely that many could have survived the storm. First impressions were fortunately incorrect in this case, and while 11 persons died in the Greensburg tornado, that is far less than expected.

The Greensburg tornado raises the question: Just how lethal is a tornado? This question is of relevance for households as they consider options in preparing for or responding to a tornado. We have examined casualties at the storm level, and now pose the question at the household level. Specifically, we ask: what is the probability that a resident will be killed if a tornado strikes her home? Not surprisingly, the answer depends on building type and tornado strength.

To precisely answer this question would require the following information: the number of structures of various types struck by tornadoes; the F-scale rating of each tornado (and since the destructive power of winds differs within a tornado and along its path, we ideally would need the F-scale rating at each location, not just the overall rating of the tornado); the number of persons in each building when the tornado struck; and the number of persons killed or injured per building. With this information, we could calculate fatality and injury rates for permanent and mobile homes by F-scale category. Unfortunately, these data are not available, and so we will use the available information to make a best estimate of the probability of fatality or injury in a tornado.

We use the narratives in the *Storm Events* database maintained by the National Climatic Data Center (NCDC) to construct these estimates. Local NWS WFOs supply this information, but they differ widely in detail. We examined the narratives of all tornadoes with property damage between

TABLE 3.11. Fatality Rates in Mobile and Permanent Homes

	Permanent Homes	Mobile Homes
Homes Struck	7331	785
Fatalities	18	17
Fatalities/Home	.002455	.02166
Fatalities/Resident	.000882	.008472
Injuries	457**	50.1**
Injuries/Home	.06237	.06393
Injuries/Resident	.02241	.02501

Calculations based on a review of tornado event summaries in the *Storm Events* database. The location of injuries is not reported, so the totals indicated by ** are inferred as described in the text, not reported totals.

2002 and 2004 for information on the number of buildings damaged or destroyed by tornadoes. Many narratives contained no information on the total numbers of buildings damaged or destroyed and thus were not used in these calculations. Some narratives gave a total number of buildings damaged or destroyed by a tornado, while others gave numbers of building damaged by type: homes—typically permanent and mobile homes—businesses, and schools or churches. Only these last narratives could be used to estimate a probability of fatality for types of structure, while the former could be used in calculating the fatalities per building. A total of 730 county-tornado narratives contained a specific number of buildings, and 503 included the number of buildings damaged by type.[13]

The detailed information on fatalities by location allows us to tally the exact number of mobile-home and permanent-home fatalities in each of the tornadoes in this sample. We do not know, however, how many people were inside each building when the tornado struck; thus, we can calculate fatalities per home, and then divide by the average household size for mobile homes and permanent homes as reported in the 2000 Census (2.56 and 2.78 persons, respectively) to estimate fatalities per resident. This is the closest approximation of the probability of death for residents of each type of home that we can obtain using this data. Table 3.11 presents the results. Fatalities per resident are about an order of magnitude greater for residents of mobile homes compared to permanent homes. Residents of mobile homes face a significant risk (although by no means a certainty) of death, with fatalities per resident of .008, or just about a 1 in 100 risk of death. The probability of death is around 1 in 1,000 for residents of permanent homes, given damage or destruction of the home in a tornado.

TABLE 3.12. Fatality Rates in Buildings by F-Scale

F-Scale Category	Buildings Struck	Fatalities	Fatalities/ Building	Injuries	Injuries/ Building
0	669	0	0	16**	.0239
1	3148	8	.00254	145**	.0460
2	6194	27	.00436	301**	.0487
3	5491	43	.00783	661**	.1204
4	2045	9	.00440	218**	.1064
5	8425	110	.01306	943**	.1193

The location of injuries is not reported, so the totals indicated by ** are inferred as described in the text, not reported totals.

Several caveats must be kept in mind when considering these numbers. We cannot control for the number of persons at home when a tornado strikes. The average number of persons in a home will be less than the number of residents, since people spend much of their time at work, at school, or shopping.[14] The inability to account for the number of persons at home should not affect the relative difference in risk between mobile and permanent homes, but the probability of death likely exceeds fatalities per resident. The narratives also report the number of homes damaged in the tornado, not the number of homes in the path of the tornado (which would include homes that were not damaged). Although we might expect that all buildings in the path of a tornado will experience some damage, this is not always true, as tornadoes are notoriously capricious in damaging buildings along their paths; in addition, minor damage might not be reported and included in the totals. Finally, some mobile homes might have been counted as homes, especially when a narrative reports the number of homes, businesses, and schools damaged. This would produce an overestimation of the probability of death for mobile homes.[15]

We use the larger sample of narratives reporting the total number of buildings struck to calculate a probability of being killed. Because this sample is larger, we also calculate fatalities per building struck by tornado F-scale rating. Table 3.12 presents the tabulations. Since the total refers to buildings in general, this likely includes non-residences, but businesses and schools also have the potential for large numbers of fatalities at one site. We cannot distinguish between mobile homes and permanent homes, so these figures may not provide accurate measures of an individual's risk. No F5 tornadoes occurred in the three years examined, so we use six F5 tornadoes since 1997 for this category. Fatalities per building increase with F-scale, as expected, but even an F5 tornado does not result in near-certain death. There were 110 fatalities in over 8,400 buildings struck in F5 tornadoes, or one fatality per 77 buildings struck. By contrast, no fatalities occurred in buildings struck by F0 tornadoes in our sample, about 1 fatality occured per 400 buildings in F1 tornadoes, one per 230 buildings in F2 tornadoes, and one per 130 buildings in F3 tornadoes. Fatalities per building do not rise with every category, as the rate for F4 tornadoes is essentially equal to that for F2 tornadoes; however, this may be due to the small sample size here, as we have seven F4 tornado segments in our sample, and the number of damaged buildings (2,045) is less than half of the totals for F2 and F3 tornadoes. Also note that the probability of a fatality in buildings struck by F0 tornadoes is not really zero, but with fewer than 700 buildings in our sample, we simply did not observe any fatalities in a building and cannot estimate the true fatality rate for F0 tornadoes (though we know it is low).

Can the differences in fatalities per building across the F-scale account for the differences in lethality displayed in Table 3.12? The regression estimates indicate that F5 tornadoes are about 1,000 times more deadly than F1 tornadoes; however, fatalities per building are only six times greater, so F5 tornadoes are not 1,000 times deadlier at the home level. Much of the difference is due to the larger damage paths of F5 tornadoes than of F0 or F1 tornadoes (see Chapter 2). The difference may also be due to the concentration of fatalities in buildings in the paths of F5 tornadoes that actually face F5 winds: only a portion of buildings in the path of an F5 tornado will suffer F5 damage; most will face weaker winds. The inclusion of buildings in the denominator that faced more modest damage may be keeping fatalities per building struck by an F5 tornado low.

The NWS does not track the location of tornado injuries, and thus we have to infer the location of injuries to be able to estimate a probability of injury to supplement fatalities per home. We have the total number of injuries for each tornado in this sample, and thus we can calculate injuries per building by F-scale category in Table 3.12. We adjust injuries in these tornadoes by the percentage of fatalities that occurred in buildings, as not all injuries occur in buildings. A study of injuries in the May 3, 1999, Oklahoma tornado outbreak reported the location of as many injuries as possible, and found that 79% occurred in permanent homes and 8% in mobile homes (Brown et al. 2002). This outbreak may not be representative because it featured an F5 tornado (63% of the fatalities in this outbreak occurred in permanent homes), but we will nonetheless use these percentages in the absence of better data. Given that permanent homes comprised nearly 10 times as many injuries in the Brown et al. (2002) study, the imputed figures for injuries per home and injuries per resident are nearly equal for permanent and mobile homes in Table 3.11, with an injury rate of just over 2% of residents. For injuries by tornado F-scale, F0 tornadoes produce one injury per 40 buildings struck, F1 and F2 tornadoes produce one injury per 20 buildings, and F3 through F5 tornadoes produce one injury per 8 or 9 buildings. Again as with fatalities, we see that injuries per building do not vary across the F-scale as expected injuries in the tornado. The regression models indicated that expected injuries were 1,500 to more than 2,000 times greater in F5 than F0 tornadoes. Although the injury rates we have estimated here are likely to have considerable noise, our method of inferring injuries should be equally accurate across tornadoes of different F-scales. The differences in total injuries again must stem more from the size of the damage path than from injuries per building.

Our estimates of the likelihood of a fatality or injury when a tornado strikes a home differ from other estimates used in research. The Multihazard Mitigation Council (2005) study of FEMA mitigation projects used death and injury rates estimated by the ATC and approved for use by FEMA. These rates are not directly comparable to those we have calculated, since the ATC/FEMA rates are based on the level of damage to a structure. The probability of death in a destroyed building is .20 and the probability of injury is .80; thus the ATC/FEMA rates assume that everyone in a destroyed building is killed or injured. Yet our analysis indicates that less than 1% of residents in a building in the path of an F5 tornado would be killed and about 5% injured. Granted, many of these buildings will not be destroyed, but our analysis simply cannot support the ATC/FEMA figures. Other available evidence also indicates that casualty rates will fall well short of the ATC/FEMA rates. A total of 2,314 homes across Oklahoma were destroyed in the May 3, 1999, tornado outbreak, and 35 fatalities in the outbreak occurred in permanent and mobile homes. Assuming that all of these fatalities occurred in the destroyed homes, there were .015 fatalities per home, with an implied probability of death of about .006. Hammer and Schmidlin (2002) interviewed survivors of this tornado, and only 1 out of 100 persons at the respondents' homes when the tornado struck was killed. Two tornadoes on February 2, 2007, in Lake County, Florida, destroyed 115 homes; 16 of the 21 fatalities in this outbreak occurred in destroyed homes, or .13 fatalities per home or .05 fatalities per resident (given that the tornadoes struck between 3 AM and 4 AM, most residents were probably home at the time). Thirteen injuries occurred in Lake County in this outbreak, so even if all of these occurred in the destroyed homes, injuries were about .04 per resident; in this case, no more than 10% of residents were killed or injured, not 100%.

3.8. What Is the Worst-Case Casualty Scenario?

Disaster movies have always enjoyed considerable popularity. In recent years, *The Day After Tomorrow* offered a string of weather disasters related to global climate change, including powerful tornadoes devastating Los Angeles. Several television series explore worst-case disasters, including The Weather Channel's *It Could Happen Tomorrow* and the National Geographic Channel's *Ultimate Tornado.*

Realistic disaster scenarios provide a basis for planning by emergency managers and policy makers, for example regarding emergency medical and housing needs after a hurricane or tornado. In conjunction with academic advisors, in 2004 FEMA created the Hurricane Pam emergency planning exercise, which centered on a hypothetical major hurricane striking New Orleans. Pam identified the dimensions of the Hurricane Katrina disaster in 2005 reasonably accurately. Unfortunately, the federal, state, and local governments failed to prepare based on the Hurricane Pam exercise, exacerbating the human tragedy of Katrina.

What is a plausible worst-case scenario for a tornado, and more significantly, what is a reasonable way to attempt to identify such a scenario? How many casualties might we expect in a worst case? To begin to address such a question, we must consider what is meant by a "plausible" worst-case scenario. Perhaps the most frightening potential mass-casualty scenario is an F5 tornado striking a major sports venue, say a football or baseball stadium or an auto race track. A tornado with winds in excess of 200 mph striking 100,000 or more persons in a relatively exposed area, with little opportunity to move to safety in the 15 minutes of lead time typically available with a warning, could result in thousands of deaths. But in reality the potential for such an absolute worst-case scenario is quite remote. Since 1950, there have been 65 F5 tornado segments in the United States, and F5 tornadoes have occurred on 45 different days. The total damage area of F5 tornadoes since 1950 is 630 square miles. While many sports and racing venues are located in states that have experienced F5 tornadoes, the probability of an F5 tornado at any given location within a tornado-prone state during a mass spectator event is very low.

Alternatively, we can use our regression models to identify a plausible worst-case path for a tornado. We used the casualty models to analyze determinants of casualties in past tornadoes, but these models can also be used to predict fatalities and injuries for a tornado of a given F-scale rating in different counties at different times of the day. Here we will focus on F5 tornadoes. The estimated coefficients of the demographic and economic variables can be combined with the characteristics of different counties to select the most vulnerable location. To limit this analysis, we consider only counties in states that have experienced at least one F5 tornado since 1900; that is, parts of the country that have exhibited vulnerability to F5 tornadoes in the past.[16] This includes virtually all of the Plains states and Southeastern states with high historical fatality totals, although there are some notable exceptions, such as Florida and Georgia. We apply values from the 2000 census for the county variables, and we use the overnight time period, even though the latest an F5 tornado began since 1950 was 11:45 PM. We also limit our attention to months in which F5 tornadoes have occurred. Since an F5 tornado occurred in February, we can use it as the deadliest month. Finally, we consider only within-county tornadoes; that is, we do not attempt to put together contiguous counties for longer-track tornadoes. A repetition of the Tri-State tornado (mentioned earlier on page 48) would need to be considered as several county tornadoes.[17]

We apply the regression model for fatalities for tornadoes from 1986 to 2007 to the counties in states that have experienced F5 tornadoes. We calculate predicted fatalities for a tornado with a path length of 50 miles during the overnight hours on a weekend in February, using the demographic characteristics of each county. Table 3.13 lists the 20 counties with the highest predicted fatalities from this exercise. Philadelphia County (PA) leads the way with an estimated 2,313 fatalities, and three other counties have a predicted death toll in excess of 1,000. The 95% confidence interval for fatalities in Philadelphia is 1,160 to 4,612, and the upper bound is our true worst case. To put the projected 4,612 fatalities in perspective, the largest death toll from a tornado in the United States was 695 in the 1925 Tri-State Tornado, and the last tornado to kill 100 persons occurred in 1953. Whether Philadelphia even represents a valid county for an F5 tornado is questionable, since the F5 tornado that struck Pennsylvania hit the western part of the state. Table 3.13 also reports the population density of each of the 20 counties; examination reveals that population density is not the only significant driver of estimated worst-case fatalities. In Philadelphia, a population density of over

TABLE 3.13. Worst Fatalities Case Counties for an F5 Tornado

County	State	Predicted Fatalities
Philadelphia	Pennsylvania	2313
Livingston	Louisiana	1262
Ascension	Louisiana	1043
Bullitt	Kentucky	1018
Cook	Illinois	948
Johnson	Texas	829
Jefferson	Missouri	807
Kaufman	Texas	697
Boone	Kentucky	694
Hidalgo	Texas	688
Montgomery	Texas	686
Will	Illinois	685
Bastrop	Texas	685
Shelby	Alabama	664
Plaquemines	Louisiana	652
Liberty	Texas	638
Lee	Alabama	627
Denton	Texas	625
Coryell	Texas	610
Rankin	Mississippi	605

The 20 counties with the highest predicted fatalities from an overnight F5 tornado of length 50 miles in February, as generated from the Poisson model of tornado fatalities for 1986–2005 in the chapter appendix.

11,000 persons/mi² drives the estimate, but the only other county considered with a population density of over 1,000 persons/mi² is Cook County, Illinois (Chicago). In Livingston County, Louisiana, a high proportion of mobile homes (32% of the housing stock) helps produce the high fatality total. With the exception of Philadelphia County and Cook County, a relatively young housing stock also contributes to the projected fatality total.

We have selected the worst possible timing in our worst-case calculations, not the modal timing for F5 tornadoes, and this considerably affects our estimates. February is the worst month for fatalities, but only 2 of the 65 F5 tornadoes since 1950 occurred in February. May is the modal month, and 78% of F5 tornadoes occurred in April, May, or June. Late afternoon is the modal time; 86% of F5 tornadoes occurred during the early or late afternoon, and tornadoes at these times are much less deadly. And of course, only some tornadoes occur on the weekend. To illustrate the role of timing, we recalculated the predicted Philadelphia death toll with these timing adjustments. Moving the tornado from February to May reduces expected fatalities by about two-thirds, to 785. Adjusting the time of day to late afternoon reduces fatalities by more than half again, to 383. And if the tornado occurred on a weekday, expected fatalities would be 292. The other totals in Table 3.13 would be reduced proportionally, and for a late afternoon tornado on a weekday in May, only the top 7 counties would have projected fatalities in excess of 100. As we can see, the definition of "worst case" dramatically affects fatality projections.

Our worst-case scenario, however, is modest compared with the projections of Wurman et al. (2007) for an urban tornado. Wurman et al. overlaid a potential long-track F5 tornado on different U.S. metropolitan areas, using census tract data to estimate the number of residents in the path of the storm, and then projected fatalities. They showed that such a tornado could readily damage or destroy the homes of more than 100,000 persons in several major metropolitan areas. They also assumed that such a tornado would kill 10% of the residents in the tornado path, and so the 630,000 residents in the path of a tornado across the Chicago metro area would result in an estimated 63,000 fatalities. As discussed in the previous section, such a fatality rate exceeds historical tornado experience; fatalities per resident in recent F5 tornadoes, the best available measure of the actual lethality of F5 tornadoes, is only about .005, or 1/20th of the rate applied by Wurman et al. If we apply a fatality rate of 0.5% to Wurman et al.'s estimate of the number of residents in the path of the Chicago tornado, estimate fatalities would be just over 3,000, which is within the confidence interval of our worst-case estimate for Philadelphia, and about three times our estimate of 948 for Cook County.[18]

3.9. Conclusion

A handful of tornadoes account for the overwhelming majority of fatalities and injuries. These tornadoes tend to be rated F4 or F5 on the Fujita Scale and cluster on the days of large-scale tornado outbreaks. Thus, while even an F0 tornado can kill, the threat to society is concentrated in the big tornado outbreaks. Fatalities and injuries are dependent on the strength of the tornado, as proxied by the F-scale rating of the tornado. Residents of permanent homes primarily face a risk of fatal injury in F3 or stronger tornadoes. Although society cannot reduce the strength of a tornado, this nonetheless comprises valuable information to alert residents to the level of risk and to allow emergency managers and medical personnel to prepare. The time dependence of casualties reflects elevated societal vulnerability. Overnight tornadoes are likely to catch residents asleep and less likely to receive a warning in time to respond, while the infrequency of fall and winter tornadoes might lead residents to dismiss an ominous-looking thunderstorm as incapable of producing a tornado, and thus slow response. Our conclusions about vulnerability must be qualified because we have not included all of the possible determinants of casualties: specifically, tornado warnings and watches issued by the NWS. For instance, the greater vulnerability posed by off-season tornadoes might be due to a reluctance of NWS forecasters to issue tornado warnings in November or January. We will add tornado warnings to the analysis in the next chapter. This will allow us to assess the contribution of NWS efforts in reducing casualties and strengthen our understanding of the determinants of casualties.

We have applied a large data set, over 50,000 state tornadoes throughout 58 years, to statistically analyze tornado casualties. Although our data set appears quite impressive, the vast majority of tornadoes do not kill or injure anyone; less than 1,300 tornadoes resulted in one or more fatality. While society benefits because most tornadoes are not in fact killers, we are left with a small set of killer tornadoes to analyze. About 100 tornadoes accounted for almost 50% of U.S. tornado fatalities between 1950 and 2007. Unpacking exactly how the various demographic and timing factors affect casualties can be difficult given the relatively small number of high-fatality tornadoes. For instance, we have seen that the mobile-home, nighttime, and winter vulnerabilities all have a concentration in the southeastern United States. Is tornado awareness and response simply low in these states? And we can only use county-level economic and demographic control variables in our analysis, creating potential imprecision in the description of tornado paths. Our analysis also reveals the value of case studies and qualitative research. Qualitative and quantitative research are often seen as substitutes, perhaps because social scientists tend to specialize in one or the other type of research, but they can complement each other. Our large data set quantitative analysis identifies patterns of vulnerability, such as the greater lethality of nighttime and off-season tornadoes, and we can offer some informed speculation on the cause of such vulnerability, such as the difficulty of ensuring receipt of warnings at night. Yet, statistical analysis cannot confirm that this is the true nature of the vulnerability. Our understanding of societal vulnerability could be substantially advanced by qualitative research to explore the basis of some of the vulnerabilities identified here.

3.10. Summary

Our analysis has revealed several patterns in casualties of relevance for understanding and ultimately reducing the societal impact from tornadoes. First, although tornadoes are nature's most powerful storm, only a small percentage of tornadoes actually result in fatalities or injuries. Second, tornadoes have become less deadly over time, and the frequency of high-fatality tornadoes has substantially declined, even though the potential for a 100-or-more-fatality tornado still exists. Third, the oft-discussed mobile-home problem for tornadoes is real, as fatality rates are about 10 times greater than for permanent homes. Fourth, nocturnal tornadoes are significantly more dangerous than daytime tornadoes. Fifth, tornadoes during the off-season winter and fall months are significantly more deadly than spring or summer tornadoes, even though the most powerful tornadoes occur in the spring. Finally, the lethality of tornadoes varies significantly across the nation, even when controlling for tornado frequency and strength as well as a wide range of other factors. Yet, several factors that are normally associated with vulnerability—the proportion of young and old and minority residents—do not drive tornado casualties in our analysis.

3.11. Appendix

The storm-path variables were constructed using the values from the decennial censuses, with values for years between the censuses inferred by linear interpolation. The values of the path variables for tornadoes striking more than one county are the averages of the variables for each county. A weighted average based on the portion of the path length in each county would be a superior method of constructing the path variables, but we have only the total path length for the tornado, not the portion in each county, and thus no way to construct a weighted average. The variables were constructed using the 1990 and 2000 censuses and the 2006 American Community Survey for counties with these variables reported. Census values from 1990 were used for tornadoes from 1986 to 1989.

We discuss in this appendix the regression models on which our discussion of the multivariate regression results in Section 3.4 is based. First note that fatalities and injuries take on nonnegative integer values; that is, the number of persons killed in a tornado can equal 0, 1, 2, or more. Consequently, ordinary least-squares regression would not be appropriate in this case since it would not take into account the truncation of casualties at zero. Economists have applied a Poisson regression model when the dependent variable is a count variable, as in this case. The Poisson model assumes that the dependent variable y_i is drawn from a Poisson distribution with parameter λ_I, or

$$\text{Prob } (Y_i = y_i) = e^{-\lambda_i} * \lambda_i^{y_i} / y_i!, \ y_i = 0, 1, 2, \ldots$$

The parameter λ_i of the distribution is assumed to be related to the independent variables x_i in a log-linear fashion, or $\text{Ln}(\lambda_i) = \beta x_i$. The Poisson regression model assumes equivalence of the conditional mean of yi and its variance, and violation of this condition is known as overdispersion. The negative binomial regression model, a generalization of the Poisson model, is recommended when the data exhibit overdispersion. Diagnostic tests indicate that tornado injuries, but not fatalities, are overdispersed.

Table 3A.1 presents the regression models for casualties over the 1950–2007 period and Table 3A.2 presents the models for the 1986–2007 period. For generality we present both the Poisson and negative binomial models for fatalities and injuries for both data sets; however, our discussion of the results in the text draws on the Poisson model for fatalities and the negative binomial model for injuries, based on the evidence of overdispersion for injuries but not fatalities. The tables report the raw regression coefficients and standard errors. To interpret the coefficients as discussed in the text, the anti-log of the coefficient must be taken. Thus, to calculate the marginal effect of a dummy variable with coefficient β_k from the table, the percentage change in expected casualties is $100 * (\exp(\beta_k) - 1)$. The percentage change in expected casualties due to a one standard deviation increase in variable ς_k is $100 * (\exp(\beta_k * \varsigma_k) - 1)$. The independent variables in each model are as described in the text, except that we include an interaction term between length and population density, on the grounds that a tornado with a long track through a densely populated area might have an interactive effect on casualties that would not be captured by either variable separately. Note that for a set of mutually exclusive categories (F-scale categories, parts of the day,

TABLE 3A.1. Regression Analysis of Tornado Casualties, 1950–2007

| | Fatalities | | Injuries | |
	Poisson	Neg. Binom	Poisson	Neg. Binom
F1	2.45** (.240)	2.50** (.247)	2.39** (.041)	2.24** (.052)
F2	3.91** (.235)	3.94** (.244)	3.83** (.040)	3.63** (.057)
F3	5.87** (.232)	5.79** (.246)	5.35** (.039)	5.03** (.078)
F4	7.69** (.232)	7.75** (.263)	6.88** (.039)	6.52** (.130)
F5	9.19** (.234)	9.44** (.377)	8.04** (.040)	7.72** (.3398)
Density	.136** (.036)	.179** (.080)	.175** (.006)	.496** (.047)
Income	−.004* (.002)	−.017** (.004)	.023** (.0004)	.005** (.002)
Time Trend	−.006** (.001)	.011** (.003)	−.008** (.0003)	−.002* (.0014)
Overnight	.271** (.060)	.596** (.114)	.023(.014)	.233** (.064)
Morning	−.162** (.072)	−.254* (.1313)	.137** (.014)	.117** (.059)
Evening Rush	.064* (.036)	−.017(.084)	−.037** (.009)	−.154** (.043)
Late Evening	.297** (.0476)	.337** (.106)	.055** (.012)	.106* (.056)
Weekend	.172** (.033)	.171** (.071)	.036** (.008)	.044(.0379)
January	1.44** (.154)	1.05** (.244)	1.01** (.030)	1.19** (.115)
February	1.57** (.141)	1.38** (.218)	1.14** (.027)	1.18** (.107)
March	1.25** (.133)	.996** (.190)	.804** (.024)	.716** (.083)
April	1.19** (.130)	.799** (.180)	.881** (.023)	.533** (.073)
May	.959** (.131)	.416** (.182)	.456** (.024)	.0296 (.070)
June	.849** (.134)	.294(.1908)	.419** (.024)	.116 (.072)
August	1.06** (.155)	.690** (.223)	.618** (.029)	.167* (.089)
September	.804** (.1645)	.684** (.227)	.234** (.032)	.118 (.096)
October	1.11** (.162)	.893** (.231)	.742** (.030)	.419** (.101)
November	1.34** (.143)	1.01** (.207)	.9969** (.026)	.855** (.092)
December	1.11** (.152)	.832** (.241)	.968** (.029)	.806** (.115)
Length	.010** (.0005)	.019** (.002)	.012** (.0001)	.028** (.002)
Length*Density	.008** (.002)	.037** (.010)	.0087** (.0003)	.053** (.010)
Intercept	−7.74** (.268)	−7.60** (.314)	−4.63** (.046)	−4.02** (.103)
Log Likelihood	−9219	−5978	−113571	−32075
Pseudo R–Sq.	.605	.287	.623	.162

Regression coefficients with standard errors in parentheses. * and ** indicate significance at the .10 and .05 levels respectively.

months of the year), one of the dummy variable categories must be omitted for the model to be estimated. The impact of the included variables is then measured relative to a tornado in the excluded category: early afternoon for day parts, July for month, and F0 for F-scale. Tables 3A.1 and 3A.2 also indicate the statistical significance of each of the coefficient estimates at two different levels, 10% and 5%, in a two-tailed test of the null hypothesis that the coefficient is zero.

TABLE 3A.2. Regression Analysis of Tornado Casualties, 1986–2007

	Fatalities		Injuries	
	Poisson	**Neg. Binom**	**Poisson**	**Neg. Binom**
F1	3.28** (.418)	3.09** (.412)	2.54** (.055)	2.43** (.074)
F2	5.21** (.410)	4.97** (.406)	4.32** (.053)	4.17** (.092)
F3	7.02** (.408)	6.51** (.413)	5.65** (.053)	5.19** (.141)
F4	8.35** (.411)	7.94** (.447)	6.77** (.054)	6.57** (.247)
F5	10.63** (.422)	10.50** (.779)	8.37** (.059)	8.33** (.835)
Density	−.0000 (.0001)	.0002 (.0002)	.0001** (.0000)	.0004** (.0001)
Income	−.011 (.0068)	−.022* (.012)	−.0027* (.0014)	.011* (.0058)
Mobile	2.40** (.305)	1.83** (.813)	1.02** (.126)	2.32** (.519)
Rural	−1.61** (.186)	−.86* (.299)	−.936** (.042)	−.496** (.149)
Non-white	−.788** (.311)	−.314 (.535)	.574** (.070)	.392 (.2796)
Male	−1.66 (2.13)	.554 (3.56)	−6.99** (.543)	−5.54** (1.81)
Under 18	−1.43 (1.52)	−4.91* (2.79)	−7.26** (.343)	−5.41** (1.26)
Over 65	−5.31** (1.44)	−7.58** (2.38)	−7.32** (.323)	−4.84** (1.02)
Commute 30+	1.47** (.421)	1.47** (.736)	.775** (.094)	−.3873 (.3628)
No High School	2.54** (.676)	2.41** (1.16)	.607** (.156)	2.55** (.596)
College	−2.09** (.700)	−.483 (1.19)	−2.13** (.162)	−.370 (.606)
Home Age	−.016** (.005)	−.0015 (.009)	−.010** (.0013)	−.013** (.0044)
Poverty Rate	−1.04 (1.05)	−1.45 (1.87)	−.512** (.219)	−1.53* (.876)
Time Trend	.062** (.0075)	.037** (.013)	.0076** (.002)	−.0088 (.0068)
Overnight	.915** (.108)	.961** (.196)	.345** (.025)	.220** (.1116)
Morning	.187 (.138)	.0079 (.2434)	.259** (.026)	.224** (.1009)
Evening Rush	.346** (.086)	.235 (.158)	.013 (.0185)	−.159** (.0731)
Late Evening	.688** (.103)	.543** (.187)	.119** (.024)	.144 (.0955)
Weekend	.202** (.069)	.071 (.123)	.1196** (.0155)	.0739 (.0638)
January	1.10** (.329)	1.29** (.430)	.416** (.0487)	.967** (.175)
February	2.19** (.304)	1.48** (.419)	.943** (.045)	.416** (.186)
March	1.31** (.293)	1.22** (.369)	.365** (.040)	.367** (.140)
April	1.24** (.292)	1.03** (.362)	.268** (.040)	.239* (.126)
May	1.23** (.289)	.57 (.357)	.142** (.039)	−.340** (.119)
June	.306 (.324)	.206 (.391)	−.164** (.044)	−.126 (.122)
August	1.77** (.309)	1.13** (.418)	.686** (.045)	−.0903 (.1528)
September	.650* (.367)	.602(.435)	−.1908** (.055)	−.0907 (.155)
October	1.43** (.336)	1.13** (.406)	.094* (.0549)	.0621 (.158)
November	1.45** (.298)	1.25** (.371)	.446** (.041)	.367** (.140)
December	1.76** (.339)	1.45** (.463)	.4266** (.057)	.240 (.215)
Length	.006** (.002)	.018** (.005)	.0095** (.0004)	.034** (.0052)
Length*Density	.0000** (.0000)	.0000(.0000)	.0000** (.0000)	.0001** (.0000)
Intercept	−9.45** (1.54)	−8.12** (2.53)	2.74** (.351)	.901 (1.18)
Log Likelihood	−2379	−1861	−30371	−11178
Pseudo R−Sq	.595	.326	.625	.187

Regression coefficients with standard errors in parentheses. * and ** indicate significance at the .10 and .05 levels, respectively.

4

TORNADO WARNINGS: HOW DOPPLER RADAR, FALSE ALARMS, AND TORNADO WATCHES AFFECT CASUALTIES

4.1. Introduction

The National Weather Service (NWS) began issuing tornado warnings in 1953. Tornado warnings represent the core element of the nation's efforts to reduce tornado casualties. In addition to warnings, the Storm Prediction Center (SPC) issues tornado watches, and more recently began issuing convective outlooks. Watches alert residents that conditions are favorable for the development of tornadoes, while a warning means that a tornado either has been spotted or has been indicated on radar. The accuracy of tornado warnings has increased over the decades, and with the advent of Doppler weather radar, warnings can now generally be issued *before* a tornado actually touches down. In 2004, the probability of detection (or the proportion of tornadoes warned for) was .69, while the average lead time on tornado warnings was almost 13 minutes.

In this chapter we examine the effectiveness of NWS efforts to reduce tornado casualties. We begin in Section 4.1 by considering the topic of warning response, or whether people respond (or want to respond) to a tornado warning when issued. Whether residents should respond to a tornado warning may seem an absurd question. Tornadoes are nature's most powerful storms and can contain winds in excess of 200 mph, capable of twisting trees, tossing cars around like toys, and leveling homes. If a tornado is bearing down on your home, of course you will want to take cover.[1] But the question we consider is whether a resident will want to take cover *when a warning is issued*, or equivalently when she learns that there is a valid tornado warning in her area. Response to the warning as opposed to the actual tornado depends on the confidence that residents place on the warning, that is, whether a warning conveys a sufficiently high level of risk to be worth responding to. Some residents might think that tornado sirens are sounded "all the time," and yet they have never actually seen a tornado. Such an attitude is not without some basis in the truth: The warning false alarm ratio in 2004 was .74, meaning that a tornado did not occur within the warned area during the period of the warning nearly three out of four times. And historically warnings have been issued for counties, so the tornado being warned for might be 20 or more miles from a given resident's home. The probability that a tornado will strike any one home in a warned county is low, and so warning response is a non-trivial decision. We will address warning response from the viewpoint of traditional hazards warning response research, and through application of the economic model of the value of information. Some rough calculations suggest that residents may not find it worthwhile to respond to tornado warnings, and may instead want confirmation that the suspected tornado is indeed approaching their home.

We then evaluate several components of the NWS warning process. We begin in Section 4.5 by considering the impact of the installation of a national network of Doppler weather radars by the NWS in the 1990s. These WSR-88D radars were a prominent component of the modernization of the NWS, and we use the installation date of the radars to test whether Doppler radar reduced casualties. We next turn in Section 4.6 to the effect of individual tornado warnings and warning lead times on casualties, and then to the influence of false alarms on casualties in Section 4.7. Almost three out of four tornado warnings are false alarms, so like the boy who cried wolf, we might be concerned that the frequent false alarms are reducing the effectiveness of warnings. We use differences in the false alarm ratio across the country to test for a false alarm effect. Tornado watches constitute a different component of the warning process, and we turn in Section 4.7 to whether the issuing of a tornado watch in advance of an outbreak affects casualties. Finally, in Section 4.9, we consider an unexpected aspect of tornadoes: the cost of responding to warnings. This cost factors into the warning response decision, and we make some calculations on the total cost of warnings to the nation. The NWS issues around 3,500 county warnings a year, and the cost of responding to these warnings appears to exceed the value of lives lost in tornadoes and the amount of property damage. We also consider how Storm Based Warnings for tornadoes, adopted by the NWS in 2007, will substantially reduce the time cost of tornado warnings as they greatly reduce the area under warning.

4.2. Perspectives on Hazard Warnings and Response

We examine in this section two approaches to individual response to tornado warnings. We first discuss the model of warning response devised by natural hazards researchers, and then apply an economic model of the value of information to tornado warnings. Each approach provides insights on the warning process, and the two approaches are not necessarily in conflict. The economic model assumes that residents apply a rational self-interest perspective to warning response, though this may not accurately describe all residents all of the time.

The efficacy of a tornado warning depends on the public response. Warnings that are ignored or that generate an inadequate response are of little value. One assumption on the part of some media figures and other people is that the public is essentially irrational when it comes to warnings. Social scientists studying natural hazards over the last 20 years have developed a descriptive model of how agents respond to a hazard warning, and recent research suggests that people act in a rational way as they decide how to respond when advised of a likely hazard. Public response is summarized in a series of eight steps, outlined as follows:

Step One: The public must receive the warning. While this may seem obvious, if warnings are not communicated in a way that reaches the public, the warning does not exist.

Step Two: The public must understand the warning. For longtime residents of Tornado Alley, it is hard to imagine misunderstanding a tornado warning. But what about people who live in areas unaccustomed to tornadoes or people who have recently relocated to Tornado Alley?

Step Three: The public must believe that the warning is credible. We explore this aspect in more detail in our discussion of false alarms, but the intuition is easy to grasp. A warning is not taken seriously if people do not believe the threat is real.

Step Four: The public must confirm that the potential threat is real. Imagine that you have been warned of an impending tornado. When you look out your window and see nothing but sunny blue skies, the warning has little relevance; perhaps it is for another part of the county. On the other hand, if you look out your window and see an ominous-looking storm, you have confirmed that the threat is real and could affect you.

Step Five: The public must personalize the threat. Not only must the threat be perceived as real but it must also be local and potentially affect the neighborhood.

Step Six: The public must determine if they need to take actions to protect themselves. The unpredictability of tornadoes and the small amount of lead time usually suggest that people should take cover. But another hazard, such as an approaching hurricane, is not as obvious. Small hurricanes do little damage, particularly for people living in well-built structures.

Step Seven: The public must determine whether action to protect themselves is feasible. If someone is at home when the tornado warning is given and the home has a safe room, protective action is feasible and obvious. On the other hand, for people driving down a crowded interstate who see a tornado about to cross their path, the feasibility of protective action is more complex.

Step Eight: Finally, the public must take protective action. While the preceding steps involve the way individuals process the decision about their response to a warning, the final step is definitive. You hear the warning, you believe it is real, and you believe the tornado will affect you. Now, take the appropriate action.

These eight steps are outlined linearly, but the process itself may not be. In addition, there is much interaction among the phases, which complicates the process beyond what can be described in a simple list. In the final analysis, the lesson for policy makers charged with warning the public of impending hazards is to take into account the various social and cultural factors that may influence the public's response when issuing warnings.

Economics values information based on the decisions people make using that information. Thus information is valued instrumentally, as a means to the end of better decisions. One consequence of this approach is that information that does not allow people to make better decisions has no value. Tornado warnings allow people to decide to take shelter. The exact location of the shelter and the details of protective actions do not affect the structure of the decision problem, although the value of sheltering will depend on the cost and effectiveness of the protective action. The economic approach assumes that people make optimal decisions both with and without infor-

Warning Status	State of World	
	Tornado Occurs	No Tornado
Warning Issued	Verified Warning	False Alarm
No Warning Issued	Unwarned Tornado	No Event

FIGURE 4.1. A tornado warning and response matrix

mation. Clearly this assumption is unrealistically optimistic, but methodologically it helps ensure that we value only the newly available information or warning. For instance, if we assumed people did not respond optimally to the information available to them without a warning, and then assumed that they did respond optimally when they received a warning, we would conflate the value of sheltering and the value of the tornado warning. Considering the best decisions possible based on tornado warnings allows us to estimate the potential value of information or warnings, even if this value is not always attained. Often, improved information is supplied together with education that helps people make better decisions, and so we might observe improved quality of decision making together with improved information. In theory, we can distinguish between the effect of education and the effect of information, even if unpacking the marginal effects proves to be quite difficult in practice.

A tornado warning represents a binary information signal from the NWS to residents. That is, the NWS sends one of two signals to residents during severe weather: by issuing a warning or not. A warning provides imperfect or noisy information to the resident regarding the likelihood of a tornado; warnings have the potential for error, and residents recognize this potential. In fact, we can distinguish two types of error in such a binary warning or forecast: The NWS could issue a county warning and a tornado could fail to touch down, or the NWS could choose not to issue a warning and a tornado could still occur. The value of a tornado warning in the economic or expected utility model depends on the quality of information contained in the warning signal or the skill of the forecast or warning (the error probabilities); the action the receiver takes upon receiving the signal; and the action the receiver would have taken if no information was available (Laffont 1989).

With a binary warning or forecast, we can apply a 2 × 2 matrix framework to this situation, as illustrated in Figure 4.1.[2] According to the economic model of information, a resident will recognize that tornado warnings are not perfect signals of an impending tornado, but will use the issuing or non-issuing of a warning to revise their estimate of the likelihood of a tornado.

In the formal treatment, this amounts to Bayesian updating. If the warning were perfect, then the probability of the upper left and lower right cells in Figure 4.1 would be 1. The probabilities with which the lower left and upper right cells occur represent the noise, or errors, in the warning. Figure 4.1 illustrates that two types of errors can be made in issuing a warning. First, a warning could be issued and then a tornado not occur in the area warned; this is represented in the upper right cell and constitutes a false alarm. Second, a warning might not be issued and a tornado could still occur; this is the lower left cell, or an unwarned tornado.

Consistent with the binary content of the tornado warning, we assume that residents make a decision to either respond to a warning from the NWS by taking shelter or to disregard the warning. (We simplify here by ignoring the potential for residents to take other actions in response to a warning, like going outside to take pictures of the tornado.) The shelter option refers to whatever protective action a household takes for a tornado, including sheltering in an interior room or storm shelter, going to a neighbor's home or community shelter, or abandoning a mobile home and lying in a ditch outside. We abstract from the timing of a response to the warning, that is, whether a resident responds immediately to the warning or waits for additional information. The economic approach assumes that a household optimizes its response to each potential signal it could receive, so the people in the household would choose to take shelter or not if they receive a warning, and to take shelter or not if they do not receive a warning. Intuition suggests that the household will want to take shelter upon receiving a warning and not to take shelter when a warning is not issued, and this is a necessary condition for a binary information warning like this to be of value to the household. But such a pattern of response is not guaranteed to increase utility; whether it does depends on the quality of the information signal or warning, the improvement in the quality (added value) of information with the warning versus simple observation of conditions, the loss that can be avoided by taking shelter, and the cost of taking shelter. The economic model focuses attention on the marginal improvement possible with a warning. Absent NWS warnings, residents could rely on visual confirmation of a tornado, or the approach of ominous black or green clouds. We will focus on the decision to shelter if a warning is issued, and presume that a household does not shelter when a warning is not issued. Conceivably, a household may shelter even if a warning has not been issued if the probability that a tornado could occur unwarned is too high, or if a warning might have been issued but not transmitted (e.g., researchers may be uncertain if tornado sirens will be sounded). Plausibly, however, threatening conditions might occur too frequently to take shelter every time dark clouds approach.

A household will respond to a tornado warning when issued and received if the value of responding exceeds the cost. This choice framework allows us to understand how various factors can affect the response decision. The value of responding to the warning will depend on the error probabilities for the tornado warning message, again illustrated using Figure 4.1. The error in the top right cell of the figure is the probability that no tornado occurs in the area warned during the time of the warning, and this corresponds to the false alarm ratio (FAR) for the tornado warning. The NWS's probability of detection (POD) of a warning is the proportion of tornadoes warned for. The probability of a tornado when the NWS does not issue a warning equals 1 minus the POD. The typical household will not be as concerned about whether a tornado occurs somewhere within the warned area, but rather whether a tornado strikes their residence. Thus, not every validated tornado warning for a particular area results in a tornado striking the residence of a given household. Residents will have to adjust the probability of a tornado occurring conditional on a warning being issued with the likelihood that their residence will be struck if a tornado occurs.

Whether residents take shelter depends as well on the effectiveness of their sheltering option in reducing the potential for a fatality or injury if their residence is indeed struck by a tornado. This depends on the reduction in the probability of a fatal or nonfatal injury due to sheltering, which is the probability of injury if residents do not shelter minus the probability of injury if residents shelter. The value of sheltering will also depend on the value of avoiding a fatal or nonfatal injury, an issue we will return to. Finally, the response decision also depends on the cost of sheltering, which includes the value of time spent sheltering (which could be substantial if at work) and any other costs involved. For instance, if residents use mattresses and blankets for extra protection, the cost of sheltering would include the time and effort to restore the home to its normal condition after the warning. Note that the cost of a tornado shelter or safe room would not be included in this cost, since such a shelter is already in place before a warning is issued. The cost of response includes only those costs incurred while responding to the warning.

We are most interested in how these factors affect a household's likelihood of responding to a warning. This is an economics exercise in comparative statics, to say whether an increase in variable X makes response to a warning more or less likely. Our focus in this chapter will be on the role

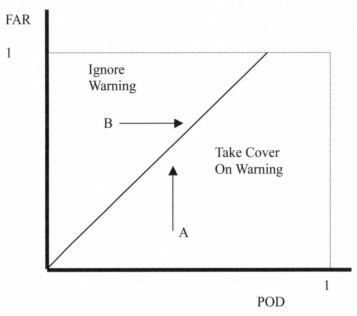

FIGURE 4.2. A resident's response to a tornado warning

of warnings, and thus we consider how the POD and FAR affect response, as illustrated in Figure 4.2.[3] The line in the figure indicates where a resident is indifferent between responding to and ignoring the warning. If the FAR and POD for tornado warnings happen to lie below the line, say at point A, the household will respond to the warning, while if the FAR and POD are at point B, the household will ignore the warning. The figure confirms the intuition that an increase in the POD, everything else equal, makes a household more likely to respond. An increase in the POD shifts the quality of the warning horizontally to the right, as indicated by the arrow from point B; a sufficiently large shift in this direction will move the household into the response region. An increase in the FAR shifts the quality of the warning vertically, as indicated by the arrow from point A, and a sufficiently large movement in this direction can move the household into the ignore region. The impact of the FAR on response illustrates the intuition of the cry wolf effect: If warnings are false alarms too often, people will start to ignore them, to their own peril when the wolf finally makes an appearance. The exact location of the line dividing the "response" and "ignore" regions depends on other parameters of the choice problem, and will differ across households. Consequently, some households might find warnings of a given quality worth responding to, while others might choose to ignore these warnings. The higher the FAR is, the larger the proportion of households in the area that ignore warnings will be, and thus tornadoes should be more deadly there. If people ignore warnings, then we will find that the issuing of warnings does not reduce casualties. This is the basis for our analysis of the effect of the FAR on casualties in Section 4.7.

We can also use this model to understand how the other parameters of the response decision behave, which provides some insight on response behavior. We now discuss the various factors introduced in this chapter in turn.

The probability of a tornado strike. Not every tornado that occurs within a warned area will strike a given resident's home. Consequently, the probability that a tornado will strike a particular residence given that a tornado occurs in the warned area affects response. Everything else being equal, an increase in this probability makes response more likely, meaning that the line separating the response/ignore regions in Figure 4.2 shifts up and to the left. The probability of a tornado strike depends on the area warned; in the extreme, a tornado warning issued for the entire state of Texas results in a much lower probability, while a warning issued just for a household's neighborhood would result in a probability near 1. For a warning area of a given size, residents may be able to refine this probability by determining the exact location and direction of movement of the tornado or pretornado circulation if this information is provided in the warning message.

The cost of sheltering. An increase in the cost of sheltering will reduce the likelihood of response. The cost of sheltering will likely vary across the day, being higher for individuals at work and possibly at night, when trying to sleep. The cost of sheltering depends on the exact activities disrupted, the ease of rescheduling these activities around the tornado warning, and the activities residents can undertake while sheltering. A resident who was reading a book when a warning was issued and has a safe room could take shelter and still continue reading during the warning, so the cost of sheltering for some residents could be quite low.

The effectiveness of sheltering. An increase in the effectiveness of sheltering makes residents more likely to respond. The effectiveness of sheltering is the reduction in the likelihood of injury, particularly severe or fatal injury. This in turn depends on the probability that residents would be killed or injured if they do not shelter, and the probability of being killed or injured if they take cover. The extreme case of the most effective take-cover option would be the resident of a mobile home who can shelter in a safe room. The probability of fatality is relatively high if the resident does not respond and close to zero if she takes cover, so the reduction in risk is maximal. On the other hand, residents who do not have what they perceive to be very effective sheltering options will be less likely to respond to a warning.

The value of fatalities and injuries avoided. Responding to a tornado warning can protect life and limb. An increase in the value residents place on safety and on avoiding injuries in a tornado will make residents more likely to shelter. We will return to the value of fatalities and injuries in Section 4.2.

Several societal factors related to tornado warning response affect more than one of the above factors, and thus we cannot make unambiguous predictions concerning the impact of these factors on warning response. For example, will residents of mobile homes be more or less likely to respond to a tornado warning? Mobile homes are vulnerable to damage in a tornado, and this increases the likelihood of severe injury or fatality if residents do not shelter, which will increase response. But residents of mobile homes do not really have a safe location within their home, and this reduces the effectiveness of taking cover, making response less likely. Therefore, in order to access a safe place to shelter, residents will have to leave their homes, which translates into a higher cost of responding to warnings. As another example, consider how income affects response. Safety is a normal good (a normal good is that for which price increases when income increases), so households with higher incomes will tend to place a greater value on avoiding fatalities and injuries, making response more likely. But the value of time is also higher for higher-income residents, which increases the cost of sheltering. Income also affects sheltering by residents of mobile homes, since the median income of mobile-home households is only about 70% of the national median.

Figure 4.2 provides a perspective on previous tests of the false alarm effect (Dow and Cutter 1998). We would expect that residents would revise upward their estimate of the FAR following a missed warning. But for a dichotomous warning like for tornadoes (or hurricanes), residents will display a discontinuous response; a resident will begin responding only when her threshold for the FAR is exceeded. Residents could revise upward their estimate of the FAR following a false alarm and still be below the frontier in Figure 4.2, and therefore continue to respond to warnings. Of course, the threshold in Figure 4.2 will differ across households, so there will always be some residents on the margin who begin to ignore warnings after a false alarm. But a test of the false alarm effect needs to control for the history of warnings for the given hazard in that region, because one recent false alarm may not push many residents across the response threshold.[4]

Tornado and severe thunderstorm watches also convey information to residents. The link between information and action is less direct for watches than for warnings. Watches are typically issued several hours in advance of the first tornado of an outbreak and can extend over half of a state or more. Tornado watches signal too diffuse a threat for residents to be expected to take precautions when a watch is issued, and since taking shelter requires only seconds or perhaps minutes, residents do not have to begin sheltering when a watch is issued. The value of a tornado watch is to alert residents to the potential for a tornado, and thus the potential for a tornado warning to be issued later. Residents under a tornado watch might be alert for a tornado warning, or if outdoors, might begin to move to shelter if they later see storms approach. Tornado watches may not affect casualties directly, but instead increase the effectiveness of tornado warnings.

4.3. What Is the Value of Responding to a Tornado Warning?

Will residents respond to a tornado warning? The economic model merely offers a framework to analyze the response decision and does not prescribe the values, and thus cannot conclusively tell us whether any given household will respond. However, we can make some illustrative calculations by applying some plausible values to the parameters of the decision.

We begin with the probability that a tornado will strike a residence given that a tornado warning is issued. This probability is the product of two components. The first is the probability that a tornado strikes the county after the issuing of a warning equals one minus the false alarm ratio, and between 1986 and 2004 the national FAR for NWS tornado warnings was .763.[5] The second is the probability that a tornado strikes a particular residence conditional on a tornado striking the warned county. We can approximate this probability with the area of the typical tornado damage path divided by the area of the warned county. This probability could be refined up or down if the resident had more information about the exact location of the actual or potential tornado, but since tornado warnings were issued for entire counties until 2007, the area of the warned county is the proper divisor. The mean damage area for tornadoes between 1990 and 2002 was .305 mi^2 (calculated by the authors from the Storm Prediction Center's national tornado archive), while the area of the county or counties struck was 1,150 mi^2. The probability that any given location suffers damage when the county is struck by a tornado is

.000364, or one in 3,000; thus the probability of tornado damage at a location given a county warning is $8.62*10^{-5}$, a little less than one in 12,000.

We turn next to the probabilities of being killed or injured given tornado damage if residents do or do not take shelter. Estimation of these probabilities requires information on the number of persons in single family homes struck by tornadoes, whether they took cover or not, and the number of fatalities and injuries by shelter status for a sample of tornadoes. Unfortunately, this information is not available (see Ashley 2007 on this point), and thus the probabilities on the effectiveness of sheltering cannot be calculated. Instead, we will use the estimates of fatalities and injuries per permanent and mobile home struck by a tornado presented in Chapter 3. The probabilities of a fatality or injury if a tornado strikes a permanent home are .000863 and .0403, respectively. Ideally, we need an estimate of the reduction in the probability of being killed or injured due to taking shelter. Fatalities and injuries per resident are really weighted averages of the probabilities of trauma for residents who do or do not take cover. In a sufficiently powerful tornado, taking shelter in an interior closet or bathroom may not have the same effect on the probability of being killed (the entire home could be blown away) as in a weaker tornado.

To compare the benefits and costs of taking shelter, we must apply a monetary value to the value of injuries and fatalities. Money is used in the economy as a unit of account, a common metric that can be used to measure other values; and at least as economists see it, placing a monetary value on life and limb does not in any way reduce the sanctity of life. Money is not the value in itself, but is simply a way to compare different values. Throughout their lives, people make various trade-offs between a risk of death or injury, and money or other values like time. Economists and risk analysts have developed the concept of the value of a statistical life (VSL) and value of a statistical injury (VSI) to describe the values elicited in such trade-offs. The term "statistical life" is used to remind us that the trade-offs involve small probabilities of death or injury, not a certainty. The potential death of an identified individual is a very different case, and the VSL is not appropriate to apply in these cases. In addition, the value of a statistical life does not imply that lives are for sale in any way. People make many fatal trade-offs in everyday life, like failing to make sure they have batteries in home smoke detectors, speeding while driving, or not being careful in jobs around the house. When people decide whether or not to buy optional safety features on a new car, they weigh money against a probability of death or injury. The VSL attempts to use values revealed in such trade-offs as a guide to understanding risky trade-offs in markets and for public policy purposes. Many government policies can reduce probabilities of death or injury for citizens at a cost, and the use of the VSL or VSI for these trade-offs suggests that government should use the same values as individuals do in these decisions.

Numerous studies have estimated the value of a statistical life in various marketplace trade-offs. Published studies have found values in the range of $1 million to $10 million (Viscusi, Vernon, and Harrington 2000), and the methodology is widely accepted by economists as an appropriate means of valuing life in risky situations. Different people will place higher or lower values on these small risks of death or injury, and in applying "a" value of life we are simply using a representative value. Economists have estimated the value of a statistical life implicit in various trade-offs, often focusing on wage premiums for risky jobs in the labor market (for surveys of this literature see Viscusi and Aldy 2003; Viscusi 2004). The Environmental Protection Agency (EPA) employed values of statistical lives and injuries in a cost-benefit analysis of the Clean Air Act. The Agency used a value of a statistical life of $4.8 million (in 1990 dollars) based on a meta-analysis of dozens of published studies (EPA 1997). Adjusting for inflation produces a value of $7.6 million in 2007 dollars, which we will apply throughout this study.

A VSI depends on the severity of injuries, and only a few studies of tornado injuries are available. Previous estimates of the VSI have been in the range of $30,000 to $50,000 (Viscusi 1993), and the EPA applied a variety of market values for different types of ailments resulting from air pollution. To apply one value for tornado injuries would require information on the distribution of the severity of injuries, and relatively little evidence is available on the distribution of tornado injuries. Some available epidemiological studies suggest that overall, tornado injuries are not very severe. Brown et al. (2002), for example, found that 76% of injuries in the May 3, 1999, Oklahoma tornado outbreak did not require hospitalization, and when required, hospital stays averaged seven days. Carter et al. (1989) found that 83% of injuries in the May 31, 1985, Ontario, Canada, tornado outbreak were minor, with serious injuries requiring an average hospital stay of 12.5 days. Based on these studies and the values of statistical injuries reported in economics literature, Simmons and Sutter (2006) valued tornado injuries at 1% of the value of a statistical life, or $76,000 in this case.[6]

Halting one's daily activities to take cover during a tornado warning is costly. All activities, including leisure activities, require time, and thus time spent sheltering has an opportunity cost for residents. Economists use the term "utility" to represent the value or satisfaction people get from consumption or other activities, and the cost of sheltering is the reduction in utility residents experience when sheltering instead of continuing their activites. In some cases the opportunity cost will be monetary, but in many cases it will not. Regardless of whether the cost is explicitly monetary, residents suffer a loss of utility when sheltering. That residents willingly incur this cost to reduce the chance of being killed or injured does not eliminate the cost.

The hourly wage represents a convenient, practical way to value time saved for benefit cost analysis, with leisure valued relative to working time. We lay out four cases here. In the first, the hourly wage seems most applicable as the opportunity cost of time spent under warnings. Even if workers are not docked for time spent sheltering, the disruption of productive activities is costly to society, and this can be valued at an average wage. In the second case, if people can choose the number of hours they work, then they will work until the utility from the money earned from working an extra hour equals the utility from an extra hour to devote to leisure activities (including sleeping). Thus some economists argue that leisure hours could be valued at the hourly wage, as this represents the opportunity cost of time (see the discussion in Smith et al. 1983). In the third case, people are not working at all, and thus the wage does not seem to represent an opportunity cost. Yet these persons could choose to work, and their potential earnings represent an opportunity cost. In the final case we have individuals who would not be working under any circumstances, and thus the wage seems an inappropriate opportunity cost for their time. However, we must acknowledge that the young, old, and others who do not work still consider their time to be valuable, and thus incur a cost when spending time sheltering as opposed to continuing their daily activities.

Economists have addressed the question of valuing leisure time. The valuation is complicated by the potential for multiple constraints; that is, the value of leisure time can depend on constraints on the leisure activity (e.g., does it require daylight). Smith, Desvousges, and McGivney (1983, p. 264) conclude that "the opportunity cost of time is best treated as a nonlinear function of time." For tornado warnings, though, the opportunity cost of responding to a warning may in some cases be low, essentially zero. For example, residents may be able to undertake leisure activities like reading a book or magazine while sheltering in a basement or safe room, so in fact these may be the activities "disrupted" by the warning. But there are other leisure activities whose disruption would be quite costly, such as a high school graduation ceremony.

The only practical approach to valuing time for many individuals is to use some fraction of the average wage rate. If employed persons choose how many hours to work (admittedly a debatable assumption for individuals, but certainly valid for individuals in the aggregate), the hourly wage represents the opportunity cost, so we will use the average hourly wage to value all of the time of persons in the labor force. The value of time for persons not in the labor force, including children, must be valued at less than the hourly wage. Cesario (1976) argues that based on available evidence a value of between 25% and 50% of the hourly wage could be applied, and recommends a value of one-third of the hourly wage, which we will use for persons not in the labor force. Our sensitivity analysis will apply a lower value of leisure in the calculation of the value of time.

We use the average civilian, non-farm hourly wage of $17.42 in 2007 to value time under warnings.[7] We apply a value of one-third this amount, $5.81, for persons who are not employed. Thus we use a weighted average value of time, with weights based on the proportion of the population employed. In 2007, 48% of the U.S. population was employed, which yields an average value of time of $11.38 (= .48*17.42 + .52*5.81).

4.4. Analysis: The Expected Benefit of Taking Shelter

We can now provide a rough comparison of the benefits and costs of responding to a county-based tornado warning. Table 4.1 summarizes the values we are applying for each component of the calculation. Figure 4.3 displays the estimated values of response for permanent-home and mobile-home residents, as well as the common cost. The value of responding to the warning is $0.73 for residents of permanent homes and $5.72 for residents of mobile homes, while the cost of taking shelter, based on the market value of time, is $7.74. Thus, the expected benefits appear to be less than the cost, and by an order of magnitude for residents of permanent homes. Of course, some of the component parameter estimates we use here are imprecise, so we cannot rule out the possibility that the benefit could exceed the cost, especially for residents of mobile homes. Also, the value and cost of taking cover are subjective and will differ across individuals, and so for some individuals the benefit may exceed the cost. Our sample calculations are for a representative or average resident.

Why is the value of responding to a warning so low? A first guess might be that our calculations are the result of a false alarm effect. The FAR enters into our calculations in the probability a tornado strikes a county when a warning is issued, and the effect of the FAR is to multiply the value of sheltering by (1 − FAR). Thus, if the FAR were reduced to zero, the value of

TABLE 4.1. Components of the Value of a Tornado Warning Calculation

Component	Value	Source
Probability of a tornado within warned county	.237	NOAA Tornado Warning Verification Records
Probability of tornado damage at a location within a county	.000364	SPC Tornado Archive, County Land Areas
Probability of fatality, injury if shelter not taken, permanent homes	.000882 .0224	See Chapter 3
Probability of fatality, injury if shelter not taken, mobile homes	.00847 .0250	See Chapter 3
Value of a statistical life	$7.6 million	EPA (1997, 2000)
Value of a statistical injury	$76,000	Authors' determination, based on values of injuries, tornado injuries
Length of tornado warning	40.8 minutes	NOAA Tornado Warning Verification Records
Value of time under warning	$11.38/hour	Bureau of Labor Statistics and Sutter and Erickson (2010)

sheltering would increase by a factor of about four, and yet would still be less than $3 for permanent homes. The low value of responding to a warning also does not stem from the ineffectiveness of taking shelter if a tornado strikes the home. The value of sheltering based on the value of life and injury and the probabilities of fatality and injury is $8,400 for permanent-home residents and $66,300 for mobile homes, well in excess of the cost. Adjusting for false alarms leaves a value of $2,000 for permanent homes and $15,700 for mobile-home residents, which is still well in excess of the cost. The substantial decrease in value occurs when adjusting for the probability that an individual home is struck by a tornado given that a tornado strikes the county, .000364. It is the large size of counties relative to tornado damage paths that really drives down the value of tornado warnings.

Our application of the economic model of information provides some insight into several factors that likely affect warning response and the societal value of tornado warnings. One is the value to residents of visual confirmation of the proximity of a tornado. Consider a county with 1,000 mi^2 of land area that is a perfect square, with more than 30 miles on each side. A tornado could be on the ground in this county and still be 20 or 30 miles away from a given residence, thus posing no danger to that residence. Under these circumstances, would or should this household take shelter? The value of responding to a warning could easily increase by one and perhaps two

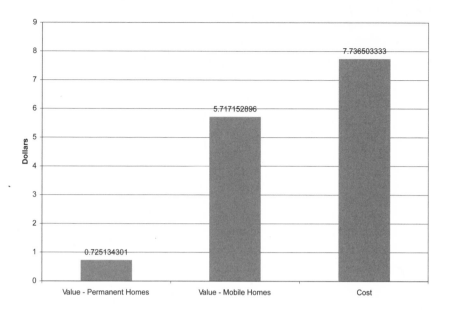

FIGURE 4.3. The value and cost of responding to county tornado warnings

orders of magnitude if residents can confirm the location and direction of movement of the tornado or wall cloud. When residents cannot confirm the proximity of the tornado, response tends to diminish. Chapter 3 identified the greater lethality of nighttime tornadoes and the correlation between state-level lethality and forest cover. Both darkness and forests interfere with visual confirmation of a tornado.

The cost of sheltering might also be substantially lower for some residents. In some cases the opportunity for sheltering might be very low. Residents might be reading a book or playing Xbox 360 when a warning is issued, and could continue to do so in a safe location. Residents could also reduce the cost of response by not sheltering for the entire duration of a warning. The average warning is in effect for 40 minutes, which might reflect the length of time for a tornado to move across a county, but as mentioned the typical warned county is quite large. Residents might only shelter while the tornado is passing through their part of the county, perhaps 10 or 15 minutes. This would reduce the cost of sheltering as estimated here by 50% to 75%.

The value of warning response can also be increased by reducing the area warned. The NWS has in fact accomplished this with the nationwide implementation in October 2007 of Storm Based Warnings (SBW) for severe weather, including tornadoes. We will return to the value of SBW later in this chapter.

4.5. Tornado Warning Efforts by the NWS

The process of providing a warning to the public of the potential hazards that tornadoes can create is a multifaceted and sophisticated exercise, though this complexity is largely invisible to the public. The contemporary NWS tornado warning process has three phases. First, when forecasters at the Storm Prediction Center note that the expected near-term meteorological conditions are ripe for the formation of thunderstorms that could spawn a tornado, a convective outlook discussing the elevated risk of severe storms is issued. The general public is often unaware of these outlooks, which are not publicized, but the outlooks alert storm spotters, emergency management staff, and broadcast meteorologists to the potential for hazardous weather over the next few hours or days. As conditions favorable for the development of a tornado begin to form, SPC forecasters will issue a tornado watch. There is greater public awareness of tornado watches than of outlooks, as a watch is broadcast across the NOAA weather alert system and through broadcast media channels. Finally, forecasters at the local NWS Weather Forecast Office (WFO) issue a warning when a tornado is indicated on radar or confirmed by spotters.

The public relies on these warnings to make safety decisions, but this has not always been the case. In the 19th century, tornadoes were a known but misunderstood weather event. The number of reported tornadoes did not exceed 200 per year until the late 1920s, which is one-fifth of the average number now known to occur annually in the United States. The event that appears to have initiated the modern tornado warning system was the Tri-State Tornado on March 18, 1925, which cut a path of destruction across Missouri, Illinois, and Indiana. Almost 700 people died in the tornado, horrifying the region and the nation. People began to realize that weather can be deadly, and public awareness of this danger contributed to the decline in the fatality rate that has continued to the current day. The NWS (previously the Weather Bureau) began issuing tornado warnings as a policy in the 1950s.

As the country prepared for the Second World War, military installations were built across the Midwest. The military soon realized that adverse weather could inflict damage to this infrastructure. The military collaborated with the Weather Bureau to create "storm spotter" networks to protect sensitive installations. Initially the concern was the effect of lightning on ammunition depots, but as time went on the spotter networks expanded their attention to other hazardous weather, including tornadoes. The storm spotter network that developed during the war continued after 1945, and was crucial in saving lives in several high-impact events during the late 1940s and early 1950s.

The first tornado warning was issued in March of 1948 by Ernest C. Fawbush and Robert C. Miller, who worked as U.S. Air Force forecasters at Tinker Air Force Base in Oklahoma City. The U.S. Weather Bureau began issuing its own warnings in the 1950s through the Severe Local Storms Forecasting Unit (SELS). The warning initiative, combined with volunteer storm spotting networks, contributed to the increase in the number of reported tornadoes during the 1950s noted in Chapter 2 to around 600 per year.

The April 11, 1965, Palm Sunday tornado outbreak was influential in the evolution of the warning process. The event began near Clinton, Iowa, and by the time the storm system pushed through Illinois, Indiana, and Michigan, almost 300 people had died in the second worst outbreak since 1950. The SELS provided warnings for the event, but the large death toll motivated the Weather Bureau to reassess the overall warning system. The post-event assessment determined that while the warnings were adequate, the public's knowledge of the warnings was deficient. As a result, the Natural Disaster Warning System (NADWARN) was formed to coordinate among various agencies with disaster and/or emergency management duties. NADWARN later became known as SKYWARN, and volunteer storm spotter networks collaborated with the NWS within this framework. The NWS began to realize that cooperation between spotters and forecasters would enhance the warning system and began to offer training programs for storm spotters. Public education and awareness programs were developed to further increase the effectiveness of warnings. The NWS now had the beginnings of what came to be known as the Integrated Warning System (IWS).

Transmitting warnings to the public and the public's response are also crucial components of an effective warning system. For years, the only direct contact the NWS had with the public was the NOAA Weather Radio network. More homes have purchased these radios in the past decade, but they still comprise a very small part of the warning dissemination process. Today, the Internet allows millions of Americans to directly access NWS warnings and forecast products, but the primary means of transmitting tornado warnings to the public remains broadcast media, as studies consistently show that television is the primary means by which people receive weather warnings (Hammer and Schmidlin 2002; Barnes et al. 2007). Whether or not an individual is aware of an impending tornado often depends on his or her access to a radio or TV in the minutes prior to the storm.

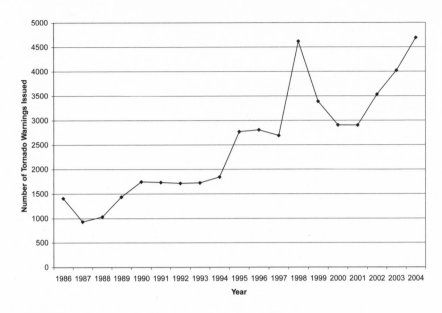

FIGURE 4.4. Tornado warnings per year issued by the National Weather Service

One major technological enhancement to the warning system was the conversion to Doppler radar. The switch to Doppler radar began in 1992, with the last of the new radars installed in 1997. In addition, the new Doppler radars were linked together for the first time into a national network. Doppler radar allows forecasters a much better view of the thunderstorm and can dramatically increase the accuracy and timeliness of warnings, as we will see shortly.

The NWS has been issuing tornado warnings since 1953, but our data for analyzing the effect of tornado warnings on casualties are from NOAA's tornado warning verification database, which dates back to 1986. We use these records in combination with our 1986–2007 data set from Chapter 3. However, we have warning verification records only through 2004, and thus our statistical analysis of warnings and false alarms on casualties will cover the years 1986–2004.

We begin by describing some overall patterns of warning performance. Figure 4.4 graphs the number of tornado warnings issued each year by the NWS. The number of yearly warnings increased by about a factor of three over these two decades, from just over 1,200 per year in the late 1980s to around 3,700 per year since 1998. The increase in warning frequency is associated with the modernization of the NWS and the installation of the NEXRAD weather radar network, which allowed forecasters to better

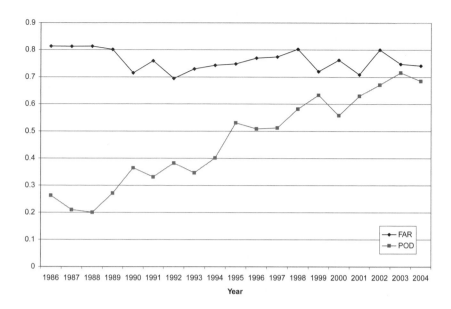

FIGURE 4.5. Tornado warning performance

observe the circulation of severe thunderstorms and warn for tornadoes. Figure 4.5 displays two metrics of warning performance, the probability of detection (POD) and the false alarm ratio (FAR). The POD is the number of tornado events warned for divided by the number of tornadoes that actually occurred, and the FAR is the number of verified tornado warnings divided by all tornado warnings. The NWS issued warnings by county during the period we are examining, so the tornado events and warnings refer to tornado segments and warnings for each county. Figure 4.5 immediately reveals the dramatic increase in the POD from between .20 and .27 between 1986 and 1989 to around .7 at the end of the period. The NWS improved from warning for about one in four tornadoes to about three in four tornadoes. The FAR remains relatively constant at between .7 and .8, which is noteworthy given the increase in total warnings over the period. At any point in time, the NWS could vary the mix of an FAR and POD in their warnings. The POD could be increased by more aggressive warnings for tornadoes, say, issuing a tornado warning for every severe thunderstorm. The absence of an increase in the FAR demonstrates that the increase in POD was not achieved by warning for every observed possible tornado circulation. Instead, the increase in POD occurred without an increase in FAR, and reflects increased skill in tornado warnings. Figure 4.6 displays the mean lead time for warnings by year. Again, a significant improvement in warning performance has occurred,

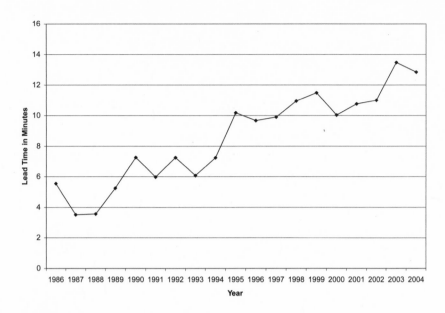

FIGURE 4.6. Tornado warning lead time

with the mean increasing from between 3 and 5 minutes in the 1980s to around 13 minutes in 2003–2004. By convention, the NWS counts the lead time for any unwarned tornado or tornadoes warned for after touchdown as zero. Thus the mean in lead time observed in Figure 4.6 is largely due to the increase in the proportion of tornadoes warned for.

4.6. Doppler Radar and Casualties

A nationwide network of Doppler weather radars was installed by the Department of Commerce as part of the modernization of the NWS in the early 1990s. Meteorologists had long been using radar to forecast the weather, and the NWS had a system of aging weather radars in place; however, these radars were not networked, which limited the ability of forecasters to use distant radars—for instance, as backup if the local radar was not operating. The WSR-88D radar employs a Doppler beam to penetrate and observe clouds and storms and had been previously used by the military. The technology was adapted for use as a weather radar in a joint effort by the Departments of Defense, Commerce, and Transportation in the late 1980s.

The NWS is organized around local Weather Forecast Offices (WFOs) that issue forecasts and weather warnings for their County Warning Areas

(CWAs). Modernization of the NWS called for a reduction in the number of local offices from over 200 to around 120, along with an increase in the professionalization of staff. Each of the newly reorganized WFO would operate its own new Doppler weather radar, which would be linked together to form a national network. Overall the NEXRAD system consists of 166 radars, 121 of which are operated by the NWS, with others located at airports and operated by the Department of Transportation or at military bases.

Doppler radars were expected to allow the NWS to issue improved tornado warnings for the nation. A benefit-cost analysis of the modernization of the NWS mentioned—but did not attempt to quantify—the number of lives that might be saved by Doppler radar (Chapman 1992). The NEXRAD radars have definitely improved the ability of NWS meteorologists to observe severe thunderstorms, and television viewers recognize the now-familiar "hook echo" within a thunderstorm that indicates the circulation of a possible tornado. We have already seen in Figures 4.5 and 4.6 that the POD and mean lead time for tornadoes increase in the mid-1990s, which corresponds with the installation of the NEXRAD radars. This in itself is suggestive, though not actually conclusive, but studies do indeed show that the installation of the Doppler radars improved the quality of tornado warnings issued by the NWS (Bieringer and Ray 1994; Polger et al. 1996; Simmons and Sutter 2005).

Improved tornado warnings are desirable, but ultimately the NWS is tasked with protecting lives, and thus the true interest of the public lies in whether Doppler radar has indeed made tornadoes less deadly. The data set we used to analyze tornado casualties can be applied to answer this question—all we need to know is exactly which tornadoes occurred after WSR-88D installation.

Our treatment variable is a Doppler dummy variable constructed using the date of WSR-88D installation at each WFO, as provided to us by NOAA's Radar Operations Center.[8] The installation date is the day when the contractor installing the radar left the site and the radar was available for forecast and warning operations. The first installation was at the Sterling, Virginia (Washington, DC) WFO on June 12, 1992, and the last installation was at the Northern Indiana WFO on August 30, 1997. Each NWS WFO issues tornado warnings for counties within its CWA. We assigned tornadoes in the SPC archive to WFOs based on the CWA of the first county in the storm path. The Doppler dummy variable equals 1 if WSR-88D radar was installed in the WFO with responsibility for the storm on or before the day of the tornado, and 0 if the tornado occurred before installation. Thus the Doppler dummy

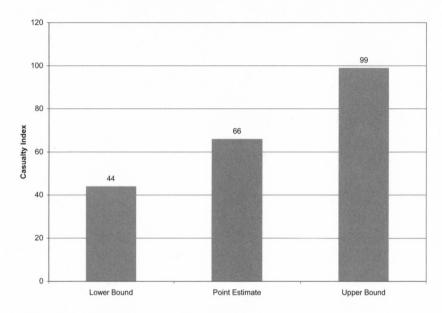

FIGURE 4.7. Doppler radar and tornado fatalities

variable equals 0 for all tornadoes in our data set occurring before June 12, 1992, and 1 for all tornadoes on or after August 30, 1997. Between these dates, the value of Doppler depends on the date of WSR-88D installation for the WFO with warning responsibility for the tornado. Overall, 70.5% of tornadoes in our data set occurred after WSR-88D installation. We have both time series and cross-sectional variation in radar. In particular, tornadoes between 1992 and 1997 allow us to determine the effect of Doppler radar as opposed to other time-varying factors on casualties.

We can use the regression model from Chapter 3 to analyze the effect of Doppler radar (and other elements of the warning process) on casualties. The Doppler radar dummy variable can be added as a control variable in the regression models. We change the econometric specification in one way from Chapter 3, by including year dummy variables in place of the integer time trend. The year variables simply equal 1 if a tornado occurred in the designated year and 0 otherwise. The year dummy variables do not impose any type of regular time trend on the relationship between time and casualties and allow greater flexibility in the specification. The year variables also control for nationwide changes over time, for example, the effect of the development of the Internet and other mobile technologies on casualties, or potential year-to-year fluctuation in casualties due to an active tornado year leading to greater public awareness of tornado risk the next year.

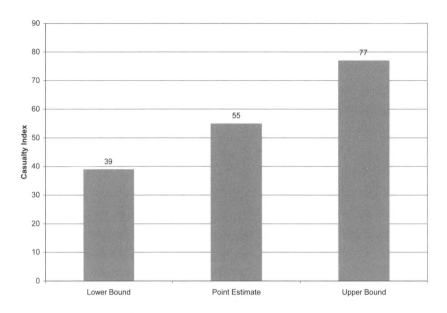

FIGURE 4.8. Doppler radar and tornado injuries

The addition of the Doppler radar variable and the year dummy variables have little effect on the estimates of the control variables already discussed in Chapter 3, so we restrict our attention to the impact of the new radar. Figures 4.7 and 4.8 depict the point estimates as well as the upper and lower bounds of the 95% confidence interval for fatalities and injuries, respectively. The figures report a casualty index with casualties for a tornado that occurred prior to Doppler radar installation normalized to 100. The point estimates—our best estimate of the impact of radar—are a 34% reduction in expected fatalities and a 45% reduction in expected injuries. Both results are statistically significant, as the upper bound of the confidence interval is less than 100 in each case. The lower bound of the confidence interval indicates that Doppler radar installation may have reduced expected casualties by more than 50%.

One of the puzzles regarding societal impacts of tornadoes mentioned in Chapter 1 was that fatalities in the six years after completion of the NEXRAD network were greater than over the six years prior to installation of the first new radar. Indeed, the increase in fatalities was substantial, from 248 in the years 1986 to 1991 to 424 in the years 1998 to 2003. This seems to suggest that Doppler radar could not have reduced casualties. However, the annual fatality total depends on the number of tornadoes, their strength, and other factors, as analyzed in Chapter 3; Doppler radar is expected to reduce fatalities only when everything else is held constant. The regression model allows us to hold the other factors constant, and when doing so we see that tornadoes in the years after installation of NEXRAD radars were more dangerous than in prior years, but ended up being less dangerous than they would have been due to the improved warnings and public response to Doppler radars.[9]

The NEXRAD network has yielded substantial returns to the nation, just based on the reduced lethality of tornadoes. Over the 10 years following completion of the network, the nation experienced 668 fatalities and 10,252 injuries. Our regression analysis indicates that these totals were 34% and 45% lower than they would have been without the Doppler radars; in other words, we estimate that NEXRAD radars prevented 348 fatalities and 8,500 injuries over this period. If we apply the values of a statistical life and injury offered earlier in this chapter, $7.6 million and $76,000, respectively, the value of tornado casualties avoided just in these years was almost $3.2 billion (in 2007 dollars).

A total of 120 of the WSR-88D radars were located at NWS WFOs in the contiguous United States (Crum, Saffle, and Wilson 1998) and are used for the forecasts and warnings evaluated here, so we consider only the cost of these radars. The radars cost $7.23 million each, and 9 radars were installed in 1992, 28 in 1993, 33 in 1994, 30 in 1995, 17 in 1996, and 3 in 1997. We date the capital cost of the Doppler radar system accordingly, resulting in a cost in 2007 dollars of $1.8 billion. To fully cost the new network we would need to include the additional cost of maintaining the WSR-88D radars, but we have no data on maintenance or other recurring costs. Still, just by using the base cost we can see that the NEXRAD network yielded net benefits of $1.4 billion based only on reduced tornado casualties.

4.7. Tornado Warnings and Casualties

We examine the direct link between warnings and casualties for a data set of more than 20,000 tornadoes nationally over 19 years. While our large data set allows us to establish the (perhaps modest) effect of warnings holding all other factors constant, it lacks measures of response to warnings at the storm level. Therefore, our investigation gauges the effectiveness of the warning process as a whole. Our analysis in this section and the remainder of the chapter is really a joint test of the effect of the warning and the response to the warning. A warning may fail to reduce casualties because the warning or extra lead time did not allow extra precautions, because the warning was not disseminated in a timely fashion, or because residents failed to respond properly to the warning. Interpretation of our results, particularly when we do not find a link between a component of the warning process and casualties, requires care, since the results may reflect poor response to a warning and not the lack of potential value of the product.

Our data set merges the SPC tornado archive with NOAA tornado warning verification records. One complication that arises here is that the NWS issues tornado warnings by county, while the SPC archive reports one entry for each state tornado segment. Consequently, several warnings might have been issued for a tornado that struck more than one county. In addition, there could be tornadoes for which a warning was issued for one county but not another, or differing lead times for warnings issued for different counties along the path. Interpretation of lead time for a second (or downstream) county would be somewhat ambiguous as well, as a warning might have been issued only a few minutes before the twister tracked into the county, and yet the tornado could have been on the ground for perhaps 30 minutes or more. To avoid some of these complications when constructing our warning variables, we only use the warnings for the first county in the storm path, since they correspond to the alert that might have been provided to residents *before* the tornado began. This allows us to avoid the confusion of determining whether a multicounty tornado was warned for or not, and if so, what lead time will be assigned to the tornado. A disadvantage of this approach, however, is that the warning variables as we have constructed them will be less representative of the potential for the typical resident in the path of the tornado to be warned for longer-track storms. We would expect our warning variables to most closely represent the effective warning status for short-path tornadoes.

We construct three types of variables to capture the warning status of a tornado. The first is a dummy variable that equals 1 if a tornado warning was issued for the tornado and 0 otherwise. Given our use of the warning for the first county in the storm path, the warning in effect dummy variable indicates whether the first county in the storm path was warned. We also use an integer variable equal to the lead time on the warning in minutes. By NWS convention, the lead time equals zero if no warning was issued for a tornado or a warning was issued after the tornado touched down. We also break down the warning lead time into categories to test for a possible variable marginal effect of lead time. We create dummy variables Lead0–5, Lead5–10, Lead11–15, Lead16–20, Lead21–30, and Lead31+, which equal 1 if

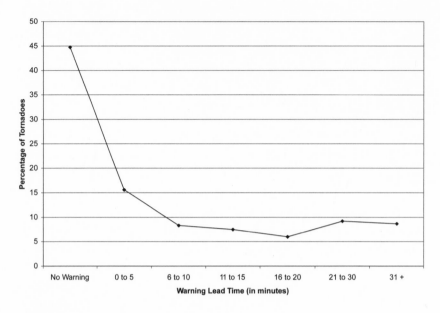

FIGURE 4.9. The distribution of tornado-warning lead times

the lead time for the warning falls in the corresponding interval and 0 otherwise. Note that only one of the warning variables is ever entered into the regression model at a time, so we will explore the impact of warned versus unwarned tornadoes, varying the integer lead time, and then warnings with lead times in the different intervals separately.

We consider first the distribution of warning lead times. Over half of the state tornado segments in our data set were warned for (55.3%), and the average lead time was 8.65 minutes, with an average of 15 minutes for warned tornadoes.[10] Figure 4.9 depicts the proportion of tornadoes in our data set with a lead time in each of the intervals. The largest proportion of tornadoes had no warning, and of the warned tornadoes, the most frequent lead-time interval was 0 to 5 minutes, with over 15% of all tornadoes. Each of the other lead-time categories has at least 5% of the tornadoes in our sample, with over 1,200 tornadoes in the category with the fewest (16 to 20 minutes). Thus we have a healthy distribution of lead times to employ in attempting to measure the marginal effect of a longer lead time on fatalities and injuries.

We begin by considering the effect of a warning on casualties. Figure 4.10 presents a casualty index for tornadoes with a warning in effect derived from the regression models. A warning reduces expected fatalities by 10% relative to a comparable tornado that is not warned for, although this estimate

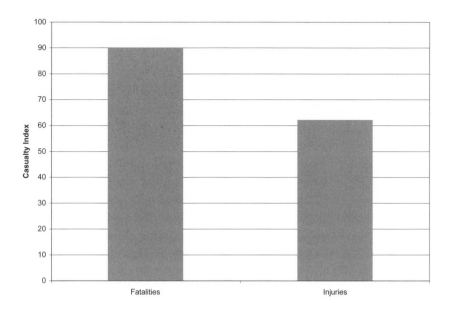

FIGURE 4.10. The effect of a tornado warning on casualties

is not statistically significant. Thus we cannot rule out the null hypothesis that a warning does not reduce fatalities at all. Warnings are much more effective for injuries, as the point estimate indicates a 38% reduction, which is statistically significant. The upper bound of the confidence interval for the injuries estimate is still a 19% reduction. The impact of warnings in the regressions runs counter to a simple comparison of casualties in warned versus unwarned tornadoes: In this data set, the 55% of tornadoes that were warned for accounted for 75% of fatalities and 68% of injuries over the period. Everything else held equal, however, warnings reduce casualties, and thus we see that the tornadoes that were warned for tended to be more lethal than the tornadoes that occurred unwarned.

Figure 4.11 depicts the effect of lead time on casualties, comparing expected fatalities and injuries for a tornado with an average lead time (8.65 minutes) against an unwarned tornado. An increase in the lead time of a warning actually increases expected fatalities, and the result is statistically significant. We can see that an average warning lead time for a tornado results in a 6% increase in expected fatalities. By contrast, a longer lead time reduces expected injuries, as intuition suggests, with an average lead time reducing injuries by 9% compared with a comparable unwarned tornado. Lead time for tornadoes is a performance measure for the NWS, and yet our

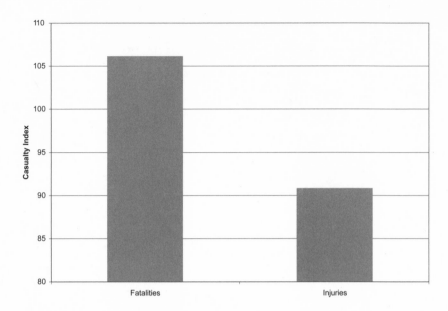

FIGURE 4.11. The effect of warning lead time on casualties. The figure illustrates the impact of issuing a warning with a mean lead time (8.65 minutes) relative to an unwarned tornado (index = 100).

results here suggest that longer lead times *increase* fatalities—a paradoxical result. When we consider the lead-time intervals, we will be better able to understand this result.

Figure 4.12 displays the effect of warnings with lead times in the different intervals, using a casualty index set to 100 for a tornado with no warning. The lead-time intervals provide insight on the relationship between warnings and fatalities. Warnings with short lead times reduce expected fatalities; the largest reduction occurs in the 6- to 10-minute range, where fatalities are reduced by 52% relative to a comparable unwarned tornado. A warning with a lead time of 5 minutes or less reduces fatalities by 19% (but is significantly different from zero at only the .10 level), and a warning in the 11- to 15-minute interval reduces fatalities by 33%. Thus, warnings can reduce fatalities. However, the figure also shows that warnings with lead times of 16 minutes or more actually increase expected fatalities, by 43% in the 16- to 20-minute range, 41% in the 21- to 30-minute range, and 8% for lead times in excess of 30 minutes (although this estimate is not statistically significant). The results are more consistent for injuries, as warnings in every interval reduce expected injuries. In fact, the greatest reduction occurs for lead times of 31 minutes or more, with a 44% reduction relative to a comparable tornado with no warn-

Dimmitt, Texas, F4 tornado on June 2, 1995, followed by VORTEX project, where the team captured images of the inner core of the tornado.

May 3, 1999, Oklahoma cluster of tornadoes produced 17 paths of destruction, which caused $1.1 billion in damage and 38 deaths.

Texas Tech University's Debris Impact Cannon simulates wind-borne debris at 100 mph to test building materials such as doors, windows, and shelters for wind storm safety.

February 2, 2007: Typical tree damage to homes in Sunshine Mobile Home Park, Lady Lake, Florida, caused by EF-3 tornadic winds of 150 mph. This mobile home park consisted of mostly older units, built in the 1970s and '80s. Because the homes were pre-1994, tie downs were sparse or widely spaced. Many of the tie downs were rusted near the ground and failed along the rusted sections.

February 2, 2007: Missile impacts a home on Norris Way, in the Village of Mallory Square, The Villages, Florida. EF-3 tornadic winds producing degrees of damage (DOD) of 3–8.

Mobile home damage due to tornadic winds.

VorTECH tornado vortex simulator at Texas Tech University uses smoke and helium bubbles to simulate the tornado's inner core.

A student stands inside the VorTECH simulator while testing the intake of air to create tornado vortices.

VorTECH: a 10-m-diameter chamber to simulate velocity and pressure profiles of tornado-like vortices at Texas Tech University.

VORTEX2 project captures data on tornado in Goshen County, Wyoming, on June 5, 2009. This tornado is the best-documented tornado to date with comprehensive data collection starting before the formation of the tornado and continuing through its demise.

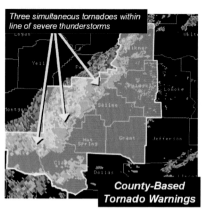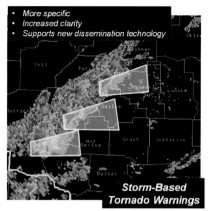

8 counties under warning
Almost 1 million people warned

70% less area covered
~600,000 fewer people warned

County-based warnings and storm-based warnings. Courtesy of National Weather Service.

VORTEX2 project team members and instrumentation arrays in June 2009.

Hill City, Kansas, tornado photographed by Dr. Ian Giammanco, WISE Center, during project WIRL on June 9, 2005.

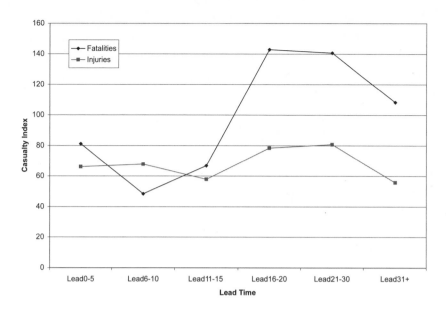

FIGURE 4.12. Warning lead-time intervals and casualties

ing, and a reduction of 42% for lead times in the 11- to 15-minute range. Lead times in the other intervals reduce injuries by between 19% and 34%.

Our analysis so far has included all tornadoes over the 1986–2004 period in one set of regressions. Conceivably, however, the value of a warning might be greater for tornadoes under certain conditions than others. For example, consider the time of day that a tornado occurs: As discussed in Chapter 3, when a tornado strikes during the overnight hours, residents may be asleep and fail to receive a warning. Our analysis consists of a joint test of the warning and warning response on casualties, and warning response at night may be impaired. Consequently, we divide our sample and estimate the regression model using the warning in effect variable for tornadoes during the day separately from tornadoes in the evening or at night. Figure 4.13 displays the results, again using a casualty index set so that an unwarned tornado has a value of 100. Warnings are more effective in reducing fatalities during the day than at night, and about equally effective in reducing injuries across the day. A warned tornado during the day results in 17% fewer fatalities and 34% fewer injuries than a comparable unwarned tornado during the day; during the evening and overnight hours, fatalities are actually 10% higher in warned than unwarned tornadoes, while a warning reduces injuries by 30% at these times. Of course, casualties are higher for all tornadoes after dark.

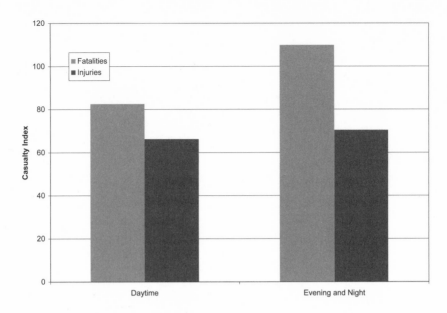

FIGURE 4.13. The effect of warnings by time of day

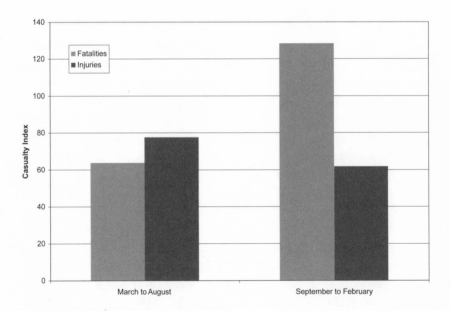

FIGURE 4.14. The effect of warnings by month

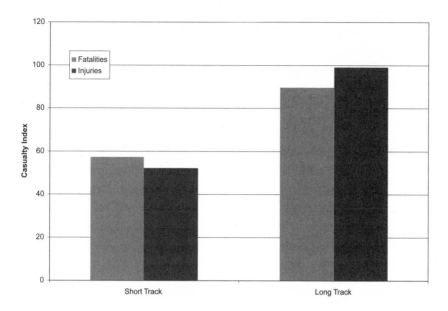

FIGURE 4.15. The effect of warnings by tornado-path length

Figure 4.14 examines the effect of warnings across the year. Here we have divided our data set into tornadoes that occur in the months of March through August, and tornadoes occurring during the off-season months of September through February. Warnings are quite effective in reducing fatalities during the peak months of tornado activity; fatalities are 36% lower for warned tornadoes than unwarned tornadoes between March and August. By contrast, tornadoes that are warned for have 28% higher fatalities during the months of September through February. Warnings reduce injuries by 22% during the peak season months and by 38% during the off-season months. We have some evidence (for fatalities) that people may respond less effectively to warnings during the fall and winter months, at least in terms of avoiding fatal injuries.

In constructing our warning variables, we applied the warning for the first county in the storm path. Our warning variables probably convey the amount of lead time less effectively for longer-track tornadoes than for short-track tornadoes. Indeed, for a tornado that tracks across several counties, the downstream counties might well have had warnings issued even if no warning was issued for the first county in the path; in this case, the tornado would still be counted as unwarned based on our warning variables. Since our warning variables will less effectively capture warning status for long-track tornadoes, Figure 4.15 shows the effect of estimating the casualties models separately for short-track and long-track tornadoes, with the sample broken at the median tornado-path length. As expected based on the above discussion, warnings are much more effective for short-track tornadoes, reducing expected fatalities by 43% and expected injuries by 48%. By contrast, a warning for a long-track tornado reduces expected fatalities by 10% and has essentially no impact on injuries. Recall from Chapter 2 that tornadoes rated higher on the F-scale have significantly longer damage paths on average, and thus long-track tornadoes will on average be stronger than short-track tornadoes. But the lethality of short-track tornadoes is substantially reduced by warnings, and this is probably our best evidence of the life-saving benefits of NWS tornado warnings.

We return now to the paradoxical positive relationship between warning lead time and fatalities. While the marginal value of an extra minute of lead time might plausibly become zero or even negative at some point, it is difficult to understand why 20 or 30 minutes of lead time on a warning should make a storm *more* deadly than no lead time. While long lead times can admittedly sometimes encourage dangerous behaviors by residents, these would need to be quite widespread in order to offset protective actions taken by residents. It is important to realize in interpreting this result that tornadoes with longer lead times are not the same as tornadoes with shorter lead times. Warning performance is better for tornadoes rated higher on the F-scale when measured by either probability of detection or mean lead time, and more dangerous tornadoes are also better warned for. Sixteen tornadoes in the data set analyzed here killed 10 or more persons, and of these tornadoes, 14 were warned for. We simply do not know how many fatalities would have resulted if the 1999 Moore, Oklahoma, or 1997 Jarrell, Texas, F5 tornadoes had occurred with no warning, or with a shorter warning. A good warning performance on the potentially deadliest tornadoes could obscure the life-saving effects of warnings. Recall also that the most powerful tornadoes tend to occur in large tornado outbreaks, and warning performance tends to be better on these days. Furthermore, our analysis comprises a joint test of the effect of warnings and response, and for several well-warned killer tornadoes, evidence suggests that the warnings may not have been well disseminated. For example, Aguirre et al. (1988) discuss how the NWS warning for the 1987 Reeves County, Texas, F4 tornado was not disseminated to the mostly Spanish-speaking immigrant community where the 30 fatalities occurred. The February 1998 Florida tornado outbreak produced two well-warned tornadoes that killed 25 and 12 persons; however,

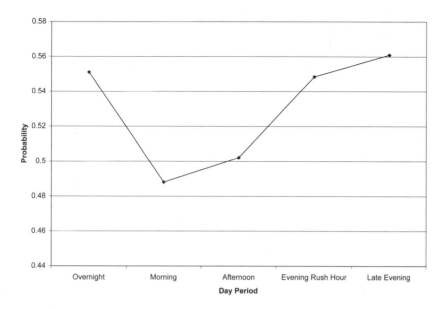

FIGURE 4.16. Probability of detection by time of day

these tornadoes occurred at night, and a lack of tornado sirens in Florida appears to have inhibited warning transmission. We therefore do not mean to imply that we think long lead times on tornado warnings are bad, in that they lead to more fatalities than would occur in the absence of a warning. Even though our data set for this analysis includes more than 20,000 tornadoes, the handful of the deadliest storms, which have also been well warned for, will drive some of our results.

Our analysis in Chapter 3 highlighted the greater lethality of tornadoes at night and during the off-season months. We now wish to consider if warning performance appears to play a part in these vulnerabilities. Note that the effects of the time of day and the month do not disappear and are essentially unchanged when we include the different warning variables in the regressions. If these vulnerabilities just happened to coincide with warning performance, the timing variables should lose their explanatory power once we control for warnings. Nonetheless, we compiled the percentage of tornadoes warned for and the mean lead time using our warning variables by our time-of-day intervals and by month. Figures 4.16 through 4.20 present the tallies. Warning performance does vary across months, with the best performances in April and May, when many of the violent tornadoes occur and when large outbreaks are more common. Warning performance is poor in July, which

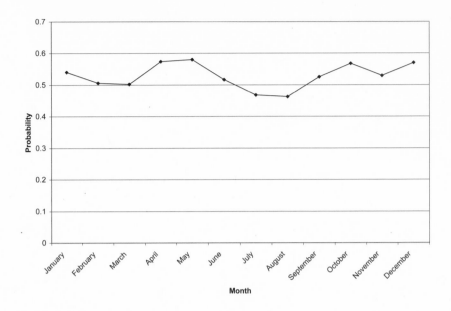

FIGURE 4.17. Probability of detection by month

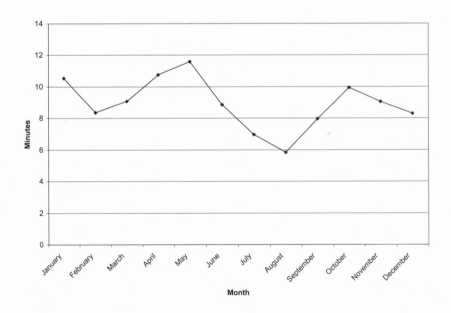

FIGURE 4.18. Average warning lead time by month

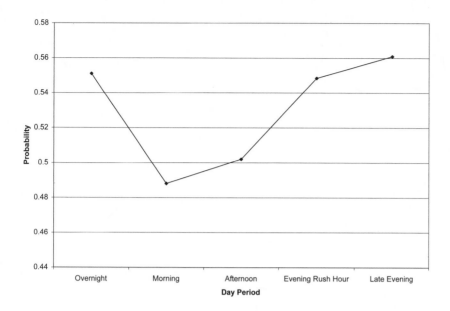

FIGURE 4.19. Probability of detection by time of day

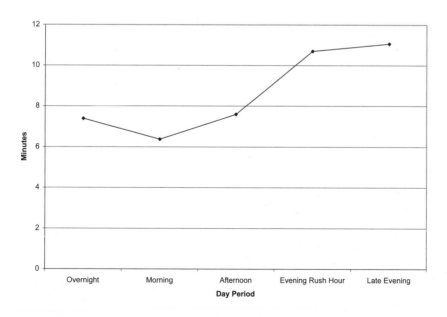

FIGURE 4.20. Average warning lead time by time of day

had the smallest fatality effect, while the POD is above average in the fall months, with average lead times. Warning performance correlates with the lethality of tornadoes across the day, but in a positive direction. The highest PODs occur in the late evening and overnight periods, when tornadoes are most deadly, so these tornadoes are being warned for on average. Average lead times are lower during the overnight hours, but the longest average lead time occurs in the late evening period. A deficiency in the quality of warnings issued by the NWS does not explain the elevated lethality of nocturnal or winter tornadoes.

4.8. False Alarms and Casualties

Theory tells us that warning performance should affect residents' subjective estimates of warning quality, but does not tell us exactly *how* residents form their perceptions. Since our data points are tornado events, the warning issued for a tornado (if any) can thus be naturally included as a control variable for the individual tornado. But false alarms are non-events, and cannot be matched with specific tornadoes. We hypothesize that residents will use recent, local warnings to try to estimate the FAR. To estimate the true FAR, residents would need to consider warnings over a certain period; warning performance varies over time, most notably with the introduction of Doppler weather radar. However, in reality, events in the distant past tend to fade from memory. Residents also use local warnings, meaning those that they directly experience, to estimate the quality of warnings: Residents of Kansas are unlikely to use warnings issued for New York, Florida, or California to evaluate the reliability of warnings.

Because what constitutes local and recent is subject to differing interpretations, we construct FARs using three geographies and two time horizons to ensure the robustness of any findings. The first geography are states; the second is NWS WFO County Warning Areas, based on the CWAs of the WFOs of the modernized NWS; and finally television markets as defined by the Nielsen Company.[11] We apply one- and two-year lags in constructing FARs for each geography, and refer to these as one- and two-year state, WFO, and TV FARs. For example, if a tornado occurred in May 2004, the one-year window would be all warnings in the relevant area between May 2003 and April 2004 (inclusive), while the two-year window would be May 2002 through April 2004 (inclusive). All tornadoes in the same geography in the same month have the same FAR.[12]

TABLE 4.2. Distribution of FAR Variables

Interval	State, 1yr	WFO, 1yr	TV, 1yr	State, 2yr	WFO, 2yr	TV, 2yr
No Warnings	1.61	3.72	5.88	0.78	1.86	2.71
1	3.43	9.65	14.29	1.76	4.44	7.31
.90–.999	6.09	7.77	6.27	3.52	6.43	5.84
.80–.899	25.40	22.95	18.27	24.89	23.01	21.08
.75–.799	16.68	12.44	12.29	18.35	17.72	16.06
.70–.749	16.73	10.49	9.83	22.25	12.63	11.86
.60–.699	23.21	15.94	17.71	22.38	20.00	21.40
.50–.599	5.08	10.48	8.78	5.81	10.77	10.49
0–.499	1.77	6.56	6.69	0.53	3.15	3.26

The numbers are the percentage of tornadoes in each category, with 1986 tornadoes excluded in the one-year variables and 1986 and 1987 tornadoes excluded in the two-year variables.

We use objective estimates of the FAR, whereas residents' sheltering decisions depend on subjective estimates of the FAR. Our use of the objective FAR does not mean that we expect residents to recall these figures from memory. Instead, our approach merely requires that residents' subjective perceptions of warning quality correlate with the objective measures. We simply require that most residents of a state with a one-year FAR of .95 should be more likely to think that "warnings are always false alarms" than would residents of a state with a one-year FAR of .25. If residents take warnings more seriously in areas where recent warning performance has been good, then response to warnings should be better, resulting in lower casualties.

Our empirical design seeks to exploit variation in the FAR across locations and time to test for a false-alarm effect. This requires sufficient variation in the FARs to affect residents' perceptions of warning quality and response to warnings. Our FAR variables all exhibit substantial variation. Table 4.2 reports the distribution of tornadoes across several intervals for each of the six FAR definitions. For the one-year FARs, between 9.5% and 21.5% of tornadoes occur with an FAR in excess of .9, while between 7% and 17% have FARs less than .6, and for each geography the FARs actually range from 0 to 1. Considerable variation exists both across and within states. To illustrate the variation in the FAR within a geography, Figure 4.21 displays the state one-year FAR over the period for Kansas, which is typical of tornado-prone states. As the figure illustrates, the FAR for Kansas ranges from .4 to .9. Consequently, our data set allows the FAR to vary over time

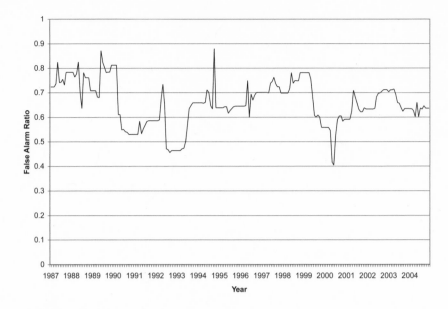

FIGURE 4.21. State one-year FAR for Kansas

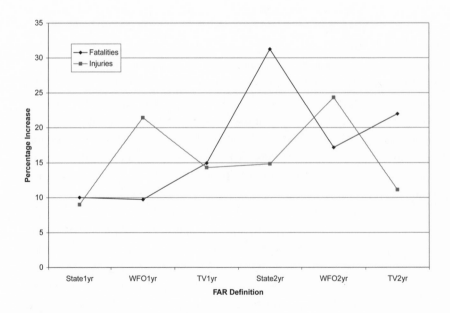

FIGURE 4.22. Effect of an increase in the FAR on casualties

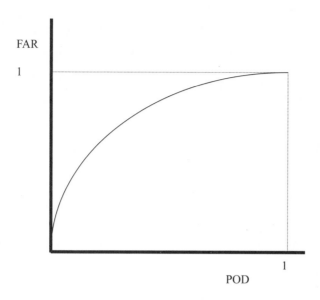

FIGURE 4.23. A tornado-warnings possibilities frontier

for tornadoes striking the same county, and the FARs do not merely reflect regional or state fixed effects.

We test for an effect of false alarms on casualties by adding the local, recent FAR variable to the casualties regression model. The model includes the warning-lead-time interval variables, and the other control variables from Chapter 3. We use each of the six differently defined FAR variables discussed above, one at a time, and are particularly interested in whether the results are consistent. Figure 4.22 displays the estimated impact of a one standard deviation increase in the respective FARs on fatalities and injuries. In all cases, the point estimates of the FAR variable are positive and statistically significant at conventional levels. The results are reasonably consistent, as a one standard deviation increase in the FAR increases expected fatalities by 10% to 31%; indeed, except for the state two-year FAR variable, the fatalities effects are all in the range of 10% to 22%. False alarms affect injuries similarly, with a one standard deviation increase in the FAR increasing expected injuries by 9% to 24%. We find strong evidence of a false alarm effect.

As discussed in Section 4.1, forecasters can make two types of errors in issuing a tornado warning: failing to warn for a tornado, and warning for a thunderstorm that does not produce a tornado. Forecasters can trade off these errors; the POD can be increased by warning for any potential tornadic circulation observed in a thunderstorm, while the FAR can be reduced by waiting for confirmation of a tornado on the ground before issuing a warning. Figure 4.23 displays a curve based on Brooks (2004) reflecting the trade-off in the warning process between the POD and FAR. The exact location of the curve will depend on current observing technology and knowledge of the meteorological conditions leading to the formation of tornadoes. In determining a warning protocol, forecasters essentially determine where on this curve to operate. To apply economics terminology, the curve in the figure is a Warning Possibilities Curve (WPC), which depicts feasible combinations of POD and FAR for tornado warnings.

The WPC represents a constraint in an optimization problem. The solution of the optimization problem would result in warnings of the greatest value, and would involve finding a tangency point between indifference curves for the POD and FAR and the WPC. If we assume that preferences regarding warnings are derived from the impact of warnings and false alarms on tornado casualties, we can use our regression analysis to determine the casualty trade-off between the POD and FAR. This exercise is complicated because the marginal effect of a warning depends on the lead time, as Figure 4.12 illustrated. We have explored the question of whether the NWS might be able to reduce tornado casualties by adjusting its warning policy (moving along the WPC), based on the casualties trade-off derived from our regression models (see Simmons and Sutter 2009). Given the shape of the WPC described by Brooks (2004), it does not appear that the NWS could reduce casualties by trading a higher POD for a higher FAR, or a lower FAR for a lower POD. In other words, the NWS appears (within the limits of confidence intervals for warnings and FAR) to be optimally issuing tornado warnings to minimize casualties.

4.9. Tornado Watches and Casualties

Before a warning is issued by an NWS WFO, the Storm Prediction Center issues a tornado watch to alert the public to the potential of tornadoes several hours later. Response to a tornado warning takes only a few moments in most cases and requires little advance preparation, so at first glance it seems that a few hours' advance notice should have little effect on casualties. But the issuing of a tornado watch alerts residents to the possibility of a warning later in the day, and thus helps to ensure receipt of the warning (which typically has only a few minutes of lead time) in time to respond. Thus watches could potentially have an impact on the effectiveness of warnings and the resulting casualties.

We explore the impact of tornado watches by using tornado-watch records supplied to us by the Storm Prediction Center. These records allow us to determine which tornadoes in our data set occurred within a valid tornado watch. We code the watch status of tornadoes based on the first county in a state tornado segment path, so if a tornado begins in a watch box and moves out of the box we code this tornado as occurring within a watch, and vice versa if a tornado begins outside but later moves into a watch box. The SPC watch records include the time at which the watch was issued, and thus we are able to calculate the lead time on the watch, in hours. In addition, we have records for severe thunderstorm (STS) watches issued by the SPC, and thus can determine if a tornado occurred within an STS watch in a similar manner as tornado watches. The expected effect of a thunderstorm watch is uncertain. On the one hand, an STS watch could alert residents of the potential for tornadoes later, similar to a tornado watch. But on the other hand, an STS watch might lull residents into a false sense of security; since a tornado watch was not issued, residents might think a later thunderstorm would be incapable of producing a tornado. Watches are thus handled as treatment variables for tornadoes, with tornado watch and STS watch dummy variables equaling 1 if a tornado began within the specified watch and 0 otherwise. These dummy variables can then be used in our casualties regression models.

We have good variation in tornadoes occurring within and outside of watch areas. Overall, 45.7% of tornadoes in our data set occurred within a tornado watch and 18.1% within an STS watch, leaving 36.2% to occur outside of any watch area. The lead time between the issuing of a tornado watch and the tornado occurring ranged up to a maximum of 12 hours, with an average of 3 hours for tornadoes that occurred within a valid watch. The maximum lead time on an STS watch was 7 hours, with an average of just under 2.5 hours for tornadoes occurring within these watches. Tornadoes in all F-scale categories occurred in each type of watch as well. Figure 4.24 displays the percentage of tornadoes by F-scale category occurring in a tornado watch or in either a tornado or an STS watch. About 60% of F0 tornadoes occurred within a watch, rising to around 70% for tornadoes rated F1 through F4, and 80% of F5 tornadoes. Stronger tornadoes are somewhat more likely to occur within a tornado watch than an STS watch; 18% of F0 and F1 tornadoes occur within an STS watch, dropping to around 10% for F3, F4, and F5 tornadoes.

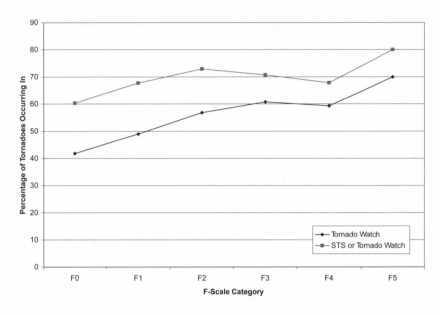

FIGURE 4.24. Tornadoes occurring in watch areas by F-scale category

Table 4.3 examines the warning performance of tornadoes by watch status. Intuition suggests that NWS forecasters would be more likely to warn for a possible tornado within a tornado watch, since the issuing of the tornado watch indicates that experts believe tornadoes are a definite possibility, and we do find evidence of this. The POD (based on our warning in effect variable for state tornadoes) is .67 for tornadoes occurring in a tornado watch box, .54 for tornadoes within an STS watch box, and .39 for tornadoes occurring in neither type of watch. But the POD certainly is not close to zero for tornadoes outside of any severe weather watch, so forecasters are not ignoring the potential for tornadoes on these days. The average lead times on warnings show similar variation, ranging from 11.6 minutes for tornadoes in tornado watches (and 17.3 minutes for tornadoes actually warned for) to 8.2 minutes (15.2 minutes) for tornadoes in STS watches, and 5.2 minutes (13.3 minutes) for tornadoes in neither type of watch.

Table 4.4 reports fatalities and injuries per tornado by watch status. Tornadoes occurring in tornado watches are most lethal on average, with fatalities per tornado almost four times greater for tornadoes in tornado watches than STS watches and about double for tornado watches compared to tornadoes in no watch area. A similar pattern is observed for injuries. The higher casualties for tornadoes that occur in tornado watches suggest

TABLE 4.3. Warning Performance by Severe Weather Watch Status

Watch Status	Probability of Detection	Average Lead Time	Average Lead Time (for warned tornadoes)
In Tornado Watch	.672	11.63	17.31
In STS Watch	.537	8.15	15.17
No Watch	.387	5.15	13.30

Average lead times are in minutes.

TABLE 4.4. Severe Weather Watches and Casualties

Watch Status	Fatalities per Tornado	Injuries per Tornado
In Tornado Watch	0.0648	1.2481
In STS Watch	0.0173	0.5540
No Watch	0.0322	0.6226

that tornado watches do not appear to be effective, but the higher casualties might instead indicate that tornadoes occurring within tornado watches are on average more dangerous than tornadoes on other days. Recall from Chapter 3 that large tornado outbreaks are responsible for a disproportionate share of casualties. SPC forecasters can likely identify days on which conditions favor large-scale tornado outbreaks and issue tornado watches on these days. Figure 4.24 showed that tornadoes rated higher on the F-scale occur more frequently in tornado watches than do weaker tornadoes, and the tornadoes occurring on these different days could differ systematically in other dimensions of lethality as well.

To more rigorously test the effect of a tornado watch, we add the different severe weather watch variables to our casualties regression models. The models include the lead-time interval dummy variables and the FAR ratio calculated for warnings issued in the state over the prior year. Figure 4.25 displays the estimated impact of a tornado watch, using an index set to 100 for a tornado occurring outside of a watch box. A tornado watch increases expected fatalities by 4% and decreases expected injuries by 14%, although the fatalities effect is not statistically significant. The bivariate comparisons in Table 4.4 found that both fatalities and injuries were much higher in tornadoes occurring within a tornado watch, so we now see that the other control variables in the regressions account for much of this difference in lethality,

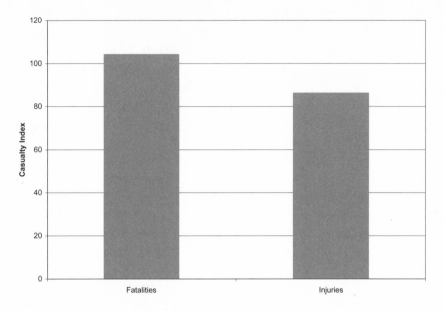

FIGURE 4.25. Tornado watches and tornado casualties

confirming the suspicion that tornadoes in tornado watches are relatively more dangerous storms. Figure 4.26 shows the results when tornadoes occurring in an STS watch are controlled for separately. In this case, tornado watches increase expected fatalities by 3% and STS watches decrease them by 7%. The impacts of the watches are reversed for injuries, with a tornado watch decreasing injuries by 9% and an STS watch increasing injuries by 18%. Thus it is difficult to make any general conclusions regarding the effect of severe weather watches on casualties.

As mentioned, the lead time on tornado watches varies substantially, with a maximum of 12 hours. Consequently the effect of a tornado watch on casualties may vary based on the lead time. To explore this we created two tornado-watch dummy variables based on the lead time; the short watch dummy variable equals 1 for tornadoes that occurred within a tornado watch but with a lead time of less than the median of 2.67 hours, and the long watch dummy variable equals 1 for tornadoes occurring within a tornado watch with a lead time greater than the median.[13] Figure 4.27 displays the results of using these dummy variables in our casualties regressions. Short watches appear to be slightly more effective than long watches. A short watch reduces expected fatalities by 2% and expected injuries by 15%, while a long watch increases fatalities by 10% and reduces injuries by 12%.

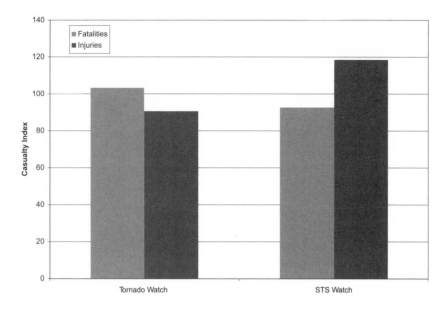

FIGURE 4.26. Tornado and severe thunderstorm watches and casualties

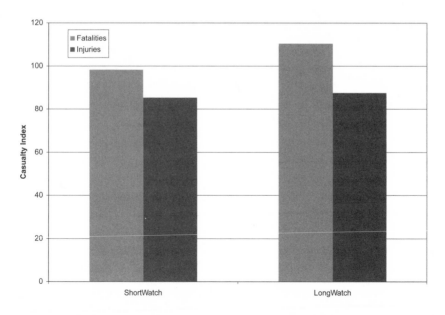

FIGURE 4.27. Short and long-lead time tornado watches and casualties

4.10. The Cost of Tornado Warnings and the Value of Storm Based Warnings

Tornado warnings reduce injuries and save lives, at least for lead times under 15 minutes. Although our simple calculation of the benefits of sheltering upon receiving a warning suggested that some residents choose not to respond, the results in Section 4.6 suggest that enough people do respond that we can detect a reduction in casualties. Still, as mentioned in Section 4.2, warning response can be costly. In this section we seek to tally up the time Americans spend under warnings, and the cost of sheltering during tornadoes. It turns out that tornado warnings impose substantial costs on the nation: in addition to residents having to take cover, manufacturing plants, retail stores, and other businesses often shut down during a warning and require extra time to resume operations.

We will focus on the time cost of tornado warnings, ignoring other potential costs. Person hours spent under warnings provide one way to tally the time cost. Person hours spent under an individual warning equals the estimated population of the warned county multiplied by the duration of the warning. We use the same county population figures as we did in constructing county population density in Chapter 3. Person hours were calculated for each county tornado warning. To illustrate the calculation, consider a tornado in Oklahoma County, Oklahoma, on October 22, 2000. A warning was issued for the county at 6:03 PM and remained in effect until 7:00 PM, for a total warning time of 57 minutes (0.95 hours). The population of Oklahoma County in the 2000 census was 660,448, so this warning was in effect for 627,000 person hours. Proceeding in this manner, we compiled a warning history for each county in the contiguous United States. Table 4.5 reports the 25 most warned counties in the country between 1986 and 2004. The most warned county is Washington County, Colorado, warned 162 times in 19 years, or about 9 times per year. Six counties in the United States were warned in excess of 100 times, and the 25th-ranked county was warned more than three and a half times a year. Table 4.5 also reports the FAR and POD for the county (based on county warnings) over the period. It's possible that the number of warnings was inflated by aggressive warning policies, which could lead to a very high FAR. However, the four most warned counties actually have an FAR below the national average, so we see that the warning total was not inflated by numerous false alarms. Indeed, only one county on the list has an FAR over 0.9.

TABLE 4.5. Most-Warned Counties in the U.S., 1986–2004

County	State	Number of Warnings	False Alarm Ratio	Probability of Detection
Washington	Colorado	162	.704	.687
Harris	Texas	149	.671	.747
Weld	Colorado	141	.610	.633
Adams	Colorado	138	.703	.590
Arapahoe	Colorado	112	.830	.568
Lincoln	Colorado	106	.830	.510
Baldwin	Alabama	94	.734	.805
Elbert	Colorado	91	.780	.643
Morgan	Colorado	87	.816	.553
Harrison	Mississippi	82	.915	.444
Brazoria	Texas	82	.817	.536
Liberty	Texas	82	.793	.759
Duval	Florida	81	.840	.609
Polk	Florida	81	.728	.304
Yazoo	Mississippi	81	.778	.833
Galveston	Texas	81	.765	.629
Cherry	Nebraska	80	.800	.760
Lincoln	Nebraska	79	.772	.632
Jefferson	Alabama	77	.857	.765
Hinds	Mississippi	77	.753	.783
Brevard	Florida	76	.882	.188
St. Johns	Florida	75	.707	.688
McLean	Illinois	74	.716	.806
Osage	Oklahoma	74	.676	.923
Logan	Colorado	73	.726	.638
Madison	Mississippi	73	.808	.778

The number of warnings issued annually by the NWS increased by a factor of three between the late 1980s and late 1990s with the installation of Doppler radars (Figure 4.4). The installation of the NEXRAD network was essentially complete by the beginning of 1996, and so we will use person hours under warnings annually between 1996 and 2004 to estimate the cost of warnings.[14] Figure 4.28 displays the annual totals of person hours under tornado warnings, which vary substantially, ranging from 164 million in 1999 to 361 million in 2004. The annual total of person hours under warnings depends on the number of warnings issued (which tracks underlying tornado

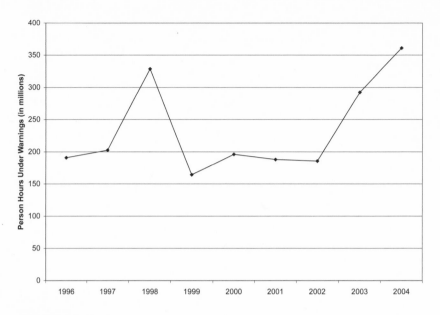

FIGURE 4.28. Person hours under tornado warnings

activity fairly closely) and the population of the counties being warned. Over the nine-year period examined here, person hours under warnings averaged 234 million per year, with the average warning in effect for 41 minutes and the average warned county having a population of 98,000.

We value time spent under warnings based on the hourly wage. Section 4.2 discussed the issues involved, and it is worth reiterating that time under warnings has an opportunity cost even if residents are not at work when a warning is issued. We use the value of time of $11.38 per hour established in Section 4.2 as a cost for sheltering.[15] We apply a value of one-third the average civilian non-farm wage ($17.42), $5.81, for persons who are not employed. Thus we use a weighted average value of time, with weights based on the proportion of the population employed. In 2007, 48% of the U.S. population was employed, which yields an average value of time of $11.38 (= .48*17.42 + .52*5.81). Applying this value of time here establishes the value of the 234 million person hours spent under warnings annually at $2.67 billion.

Time spent under warnings does not accurately measure the cost of tornado warnings, because not all time under warnings is spent sheltering. If people do not receive or choose to ignore a warning, their daily activities will not be disrupted and being under the warning will have imposed no cost on them. The number of hours actually spent sheltering each year will be less than 234 million. Some persons will shelter for only a portion of the

TABLE 4.6. Survey Studies on Tornado Warning Response

Studies Directly Asking About Response	State(s)	Response Rate	Notes
Liu et al. (1996)	Alabama	29% w/o sirens 66% with sirens	Warned county
Balluz et al. (2000)	Arkansas	46%	
Paul et al. (2003)	Missouri	89%	Communities struck by tornadoes
Tiefenbacher et al. (2001)	Wisconsin	53%	Community struck by tornado but outside of damage path
Schmidlin et al. (2009)	Georgia, Illinois, Mississippi, Oklahoma	31%	Warned county
Studies with Indirect Information			
Hodler (1982)	Michigan	48%	Storm path
Legates and Biddle (1999)	Alabama	70%	Storm path
Hammer and Schmidlin (2002)	Oklahoma	87%	Storm path

valid warning time, and others will not respond at all. Although calculating person hours under warnings is straightforward given the warning records, we have no data on the response rate to individual warnings. Furthermore, the evidence on response to tornado warnings (as opposed to other types of hazard warnings) is sparse (Sorensen 2000); surveys of residents after tornadoes provide the only available evidence. The top portion of Table 4.6 lists published survey studies of response to tornado warnings, and the reported response rates range from just under 30% to almost 90%. Based on the wording of most of the available survey questions, these rates appear to be the percentage of persons who responded at all to the warning, and no evidence is available on what portion of the valid warning time people actually spent sheltering. Consequently, we will use these reported response rates as the percentage of time under a warning spent sheltering. The bottom portion of Table 4.6 lists several additional studies that report information similar to a response rate, for instance, where residents said they were when a tornado struck, but the investigators did not ask directly if respondents sheltered. Nonetheless, locations consistent with NWS recommendations for response probably indicate that the respondents were sheltering. Although these studies are less on point, we think they are worth considering given the paucity of evidence. Based on these studies we will apply a 50% response rate, or assume that 50% of time under county warnings is actually spent sheltering. Thus time spent sheltering can be inferred to average about 117 million person hours per year, with a value of $1.33 billion.

We see that responding to warnings represents a substantial societal cost of tornadoes, and unlike the cost of hurricane warnings—evacuations—this cost has not been recognized. As we saw in Section 4.2, counties are quite large relative to the size of tornado damage paths. Based on the average damage path and warned county, we estimated that there is about a 1 in 3,000 chance of damage at any one location given that a tornado is on the ground in the warned county. County warnings essentially overwarn for tornadoes; that is, they warn many persons who are far from the actual tornado (assuming that the warning is not a false alarm). Avoiding unnecessary evacuations has been recognized as an important societal benefit of improving hurricane forecasts and warnings (Letson et al. 2007), but a similar argument applies for tornado warnings as well.

In fact, the NWS has already taken steps to reduce the cost of tornado warnings. The NWS introduced SBW for tornadoes and other types of severe weather nationally in October 2007 (NWS n.d.). SBWs are polygons based on the location of the actual or potential tornado, and are typically much smaller than an entire county. (After all, counties are arbitrary political jurisdictions and do not correspond with the area threatened by a tornado.) In tests during 2004 and 2005, SBWs reduced the area covered by tornado warnings relative to conventional county-based warnings by 70 to 75% (Looney 2006; Jacks and Ferree 2007). In the February 2008 Super Tuesday tornado outbreak, the first major test of SBWs in practice, the area warned was reduced by 61% (NWS 2009), even though two of the states affected by this outbreak, Tennessee and Kentucky, have relatively small counties.

We can use our findings of the value of time spent sheltering to place a value on the time savings with SBW. To estimate the savings on time under warnings, we can simply apply the expected reduction in the area warned, which we will assume here is 70%. The savings on time spent sheltering, however, is likely to be less than 70%, because the response to tornado warnings is likely dependent on proximity to the tornado. Some of the surveys listed in Table 4.6 involved interviews of residents living quite near the path of the tornado, so these people would still be in the polygon of an SBW. Savings on time spent sheltering with SBW will depend on the response rate to warnings in the portions of counties that would now be outside of the polygon. This response rate is likely lower because the theory of hazard warning response contends that people will seek to confirm and personalize a threat before responding (see Section 4.1), which for tornadoes often involves visual confirmation of the proximity of the tornado or wall cloud (Hodler 1982; Schmidlin et al. 2009). The 50% response rate we applied above is an average of the response rates in the portions of counties that are within and outside the new SBW polygon. We turn again to the surveys listed in Table 4.6 for guidance on what the response rates within and outside of the polygon might be. The notes for the studies indicate whether the sample was drawn from an area in close proximity to the tornado or from across the warned county when that information is available. The higher response rates do tend to be in samples in closer proximity to the tornado and thus likely inside the SBW polygon. If we take 50% to be the overall response rate across an entire warned county, we assume a response rate of 40% outside of the polygon, which implies a response rate within the polygon of about 73%.[16]

Even with these adjustments, the time savings with SBW are likely to be substantial. In our sample, an average of 234 million person hours were spent under county warnings each year, with a value of $2.67 billion. SBW will reduce person hours spent under warnings by 70%, to 70 million per year. This represents a savings of 164 million person hours spent under warnings, worth $1.9 billion annually. An estimated 117 million person hours were actually spent sheltering under county warnings annually, with a value of $1.33 billion. SBW will reduce time spent sheltering by 66 million person hours annually, with a value of $747 million, a savings of 56% of the time actually spent sheltering under county warnings.[17]

The above discussion considers only the time savings with SBW. If the warning area is reduced, the immediate question that arises is whether safety will be compromised. Because the area immediately threatened by a tornado is still warned, SBW should not compromise safety (NWS n.d.). This does not mean that tornadoes will not strike unwarned areas, however, because SBW will not allow the probability of detection to rise to 1. It does imply that any impact of SBW on casualties will depend on indirect or second order effects. A number of indirect effects are possible, including the potential for a tornado to veer out of the polygon; confusion among forecasters, emergency managers, and residents in describing the polygon area; reduction in the relaying of warnings by residents under warnings; and possibly a reduction in lead times. All of these effects might increase casualties. As discussed by Sutter and Erickson (2010), these effects seem quite modest. Additionally, SBW has the potential to save lives by increasing warning response. The expected utility model of warning response in Section 4.2 emphasized that the large size of counties relative to tornado damage paths substantially reduced the value of county-based tornado warnings. Indeed, using reasonable estimates of the parameters of this decision, it appeared that a typical resident would not find a county warning worth responding to, unless the probability that the tornado would strike the resident's home could be updated. The value of NOAA Weather Radios, local tornado sirens, and other alert systems is derived from the value of the warnings being conveyed: If the value of county warnings is low, the value of alert systems to convey these warnings will also be low. SBWs refine the county warning, producing an inherently more valuable warning signal that more people will want to respond to, and as a result will increase the value of emergency alert systems, assuming the alert system is refined enough to use the coordinates of the SBW polygon. The potential for improved response suggests that SBWs could easily lead to an eventual reduction in tornado casualties.

4.11. Summary

The first tornado warnings were issued over 60 years ago, and over this period of time casualties from tornadoes have declined significantly. Many factors have contributed to this decline in casualties, and in this chapter we have examined the impact of warnings on casualties. Of course the warnings issued by the NWS are part of a warning process that includes the observation of storms (weather radars, storm spotters), transmission of warnings via television, radio and other media, and education and response on the part of the public. Our analysis shows the effect of this overall warning process.

We have used NWS tornado warning verification records in our analysis, and these records are available beginning in 1986. Much of the decline in the U. S. tornado fatality rate evident in Chapter 3 had already occurred by this time, and so our analysis cannot trace out exactly how warnings might have produced this decline. Nonetheless, the introduction of NEXRAD Doppler Weather Radar in the 1990s by the NWS reduced fatalities and injuries by about 40% relative to the already low levels in the late 1980s and early 1990s. Warnings with lead times up to 15 minutes reduce both fatalities and injuries, but we find no additional reduction of casualties for lead times beyond 15 minutes. We also find that tornado watches have no effect directly on casualties, although they contribute as part of the warning process to saving lives. Together these findings suggest that residents do not appear to need lengthy advance notice to protect themselves from tornadoes, at least given the traditional recommended protective actions.

We also found evidence of a false alarm effect, as a higher recent, local false alarm ratio significantly increases fatalities and injuries. Identifying a "cry wolf" syndrome is significant for forecasters. A trade-off exists between the probability of detection and false alarms, and the probability of an unwarned tornado can be reduced essentially to zero by aggressively warning for any strong thunderstorm. Such a warning strategy would exponentially increase the number of false alarms, but in the absence of a cry wolf effect, aggressive warnings should reduce casualties. We have not precisely quantified the warning–false alarm casualty trade-off, but simply knowing that the trade-off exists is significant.

We are now at a point where we can examine the effectiveness of the warnings beyond simple measures like lead time. Future research can consider the role of time of day in the warning process. Nocturnal tornadoes are much more dangerous, even when controlling for warnings and false alarms, presumably because residents are less likely to receive warnings for these storms. Technology can perhaps address this problem through warning transmission systems like NOAA weather radios, reverse 911, and use of outdoor warning sirens.

SHELTERING FROM THE STORM: EVALUATING TORNADO SHELTERS AS A MITIGATION INVESTMENT

5.1. Introduction

On May 27, 1997, an F5 tornado struck the town of Jarrell, Texas, near Austin. The slow-moving tornado devastated an entire subdivision, wiping homes clean off their foundations. The casualty figures, 27 dead and 12 injured, indicate this tornado's deadly power: More than two out of every three casualties were deaths. Tornadoes generally injure more persons than they kill: Since 1950, tornadoes have injured 81,000 and killed 4,900, or about 16 injuries per fatality. A tornado with more fatalities than injures is rare, and the difference of 15 more fatalities than injuries in the Jarrell tornado is the largest difference in any tornado since 1900. There was almost no escape for the unfortunate residents caught in the path of this tornado.

Tornadoes are nature's most powerful storms, but the effect of their destructive power can be mitigated with the proper shelter. Consider the following examples:

The May 3, 1999 F5 tornado which struck the Oklahoma City area devastated entire neighborhoods. One family in Midwest City survived the storm in their walk-in closet which had been specially designed to protect against tornadoes, with 12-inch reinforced concrete walls, a concrete roof, and a steel door. Beth Bartlett had instructed the builder to reinforce the room, and she and her mother, two dogs, and two cats all survived without a scratch even though the rest of her home was destroyed and five people died in her neighborhood.[1]

An F4 tornado struck Woodford County, Illinois in July 2004 and destroyed the Parsons Manufacturing Co.'s facility. None of the 150 workers or one job applicant at the plant at the time was injured, due to three above-ground storm shelters built in the plant. Company owner Bob Parsons said, "There's no doubt about it. They [the employees] wouldn't be here today if it weren't for the shelters." The company also had a NOAA Weather Radio and conducted regular tornado drills so employees would know where to go to take cover.[2]

These anecdotes demonstrate that it's possible to design buildings (or at least shelters) that can survive even the strongest tornadoes. Wind engineers at Texas Tech University have pioneered designs for "safe rooms" made with steel-reinforced cinder blocks. The Federal Emergency Management Agency (FEMA) included safe rooms and shelters as part of their National Mitigation Strategy (since abandoned), and issued technical construction standards for shelters in 1998 (FEMA 1998). FEMA later issued construction standards for community shelters for schools and other public buildings (FEMA 2000). After the May 3, 1999, Oklahoma City F5 tornado, the National Storm Shelter Association was formed to certify the quality of shelters and safe rooms built by private contractors. And FEMA and the state of Oklahoma joined forces on the Oklahoma Saferoom Initiative to offer $2,000 rebates to homeowners installing shelters or safe rooms in their homes. More than 14,000 Oklahoma families applied for the rebates. Allocated funds for the program allowed payments to more than 6,000 families.

This chapter evaluates tornado shelters as a life-saving measure, explores the emergence of a market for such shelters, and discusses the public policy issues raised by mitigation of this type. We begin by applying the economic model of decision-making under uncertainty to shelter purchase. Shelters constitute a durable investment that must be made well in advance of a tornado strike. We then present estimates of the cost-effectiveness of shelters by state, using both tornado probabilities and the incidence of fatalities and injuries. We will find that shelters do not appear to be cost-effective for permanent homes because fatalities in these homes are just too rare; on the other hand, shelters do appear to offer cost-effective protection for mobile homes, at least in the most tornado-prone states, and provide a way to address the mobile-home vulnerability discussed in Chapter 3. We also consider how shelters can provide peace of mind even if a particular building is never struck by a tornado, and how these benefits improve the value proposition of shelters. We conclude by reviewing evidence on the market effects of natural hazards and mitigation generally, as well as applications to tornadoes.

5.2. Investing in Mitigation

Economists analyze complex market interactions on the presumption that people are rational and self-interested. In formal mathematical models, it is assumed that consumers and businesses act in a rational way and pursue their self-interest to solve complex optimization problems. Rational self-interest basically means that people act to try to make themselves better off and that they have some ability to calculate what will make them better off. Often this amounts to comparing the benefits and costs of a proposed action. We apply this framework to a household's decision to purchase a tornado shelter or safe room, and assume that residents want to compare the safety and peace of mind a shelter provides against the cost of the shelter.

The cost of a tornado shelter is fairly straightforward: it's simply the financial cost of installing and maintaining the shelter itself. Many companies manufacture and install shelters on the open market, and we can use the market price (installed) as the cost. A minor complication arises because a shelter or safe room is a durable investment and provides benefits for years, while purchase and installation costs are incurred upfront. To manage this, we need to consider either annualized costs and benefits, or costs and benefits over the useful life of the shelter; we cannot mix the total benefits with annual cost, for example. We will return to this matter shortly.

The benefits of the shelter or safe room are the injuries or fatalities avoided, and the peace of mind of knowing that you and your family and guests will be safe in the event of a tornado. Peace of mind is likely a major benefit, as the probability that any one home will be struck by a tornado in the next year—or even 10 or 20 years—is quite low (Chapter 2). Even in Tornado Alley, most tornado shelters will never be struck by a tornado. However, residents will be under a tornado warning many times over the life of the shelter, and the presence of a shelter or safe room in which to take cover will help assure residents of their safety if a tornado does strike, and reduce their anxiety during these anxious moments.

We will focus primarily on the safety benefits of shelters, and then later address the peace-of-mind aspect. This is not because we consider peace of mind to be unimportant; the assurance of safety at a time of crisis and danger may be as important as actually protecting residents. However, we can use casualty totals to estimate fatalities and injuries avoided by shelters and thus evaluate the trade-off between safety and cost, while the peace-of-mind benefit is difficult to quantify with any precision.

In deciding to purchase a shelter or safe room, residents must compare the benefits—the potential fatalities or injuries avoided—with the financial cost. When making this choice, residents place a dollar value on safety, at least implicitly. As discussed in Chapter 4, economists are comfortable placing a dollar value on the prospect of injuries or death; they take the view that life and money are not incommensurable. People implicitly place a dollar value on life every time they make a trade-off between safety and money or time.

To compare the benefits and cost of shelters, we—and in theory, individual households as well—need to estimate the number of lives that can be saved and injuries avoided over the useful life of an in-home shelter. Throughout our research on tornadoes, we have used different methods to calculate the value of tornado shelters. We will continue in this vein and offer estimates of the cost per life saved by tornado shelters derived in several different ways. We do this in part because the data required to estimate the value of a shelter as we would like are not available. Since we must imperfectly estimate lives saved, using several different methods allows us to compare the variation. If all of our calculations yield reasonably similar results, we can have more confidence that our estimates of cost per life saved are close to what we would get if we had the data to make our ideal calculations of cost-effectiveness. If our several methods result in wide variation in the final estimates, then conclusions concerning cost-effectiveness must be very tentative.

A single resident would need to estimate the probability that she would be killed or injured in a tornado each year if she did or did not have a tornado shelter in order to estimate the benefits from a shelter. The reduction in probability of death and injury would then be multiplied by the value of a statistical life and statistical injury to estimate the expected annual benefits. The resident would then need to compare the present value of this stream of benefits with the cost of the shelter, which must be incurred at the time of installation. For a household of two or more people, the probabilities of fatality and injury per person would be multiplied by the household size to derive the household's benefits from the shelter.[3] It should be noted that as a general rule, human beings are impatient, meaning that we prefer immediate consumption to deferred consumption. A rational, self-interested household will choose to invest in a shelter if it evaluates the expected benefits to be greater than the cost. A household that performs this calculation and decides against purchasing a shelter is not ignoring tornado risk or failing to appreciate the value of self-protection; rather, they are deciding that the benefits are not great enough to be worth the cost, which is the alternative use of money that would have been spent on the shelter.

We use this framework to provide evidence on the cost-effectiveness of tornado shelters. Because different households place different values on safety—they have different values of a statistical life or injury—we do not select a value of life to apply in our cost-effectiveness calculations. Instead, we estimate the lives saved by a shelter, apply the cost, and calculate the average value of life (or casualty) for which the benefits of the shelter would just equal the cost. This is known as a cost per life saved or fatality avoided, and evidence on the cost-effectiveness of risk-reducing measures is often compared in this manner (Viscusi, Vernon, and Harrington 2000). We will calculate a cost per life saved (ignoring the value of injuries avoided) and a cost per casualty avoided, which combines fatalities and injuries in a way to be discussed shortly.

The economic approach is one of several that can be applied to self-protective decisions. We apply it here because it facilitates the calculation of a value of tornado shelters, but the model implies that the self-protection choice is made in a very calculating fashion. In practice, people's behavior is not as fully rational as the economic model suggests. Some deviations from the model represent reasonable applications of the approach to a world where information is costly to acquire and decision-making itself is a costly endeavor. For example, to estimate the benefits of a tornado shelter, residents need to estimate the likelihood of being killed in a tornado if they do not have a shelter. The information needed to make such an estimate is costly to assemble, and we have in fact spent years assembling the data used in the estimates in Section 5.3. Rational residents will balance the cost of acquiring more information in order to make a better estimate of the benefits against the improvement in the quality of the decision they can make with the information. To economize on information costs, people will likely make crude calculations based on readily available information: for example, recent news stories about deaths in tornadoes. Some people will overestimate the benefits in this fashion and others might underestimate the benefits, but either way the economic framework still provides a good way to understand the decision. Residents who overestimate their likelihood of being killed will also overestimate the benefits (relative to the best available information), and will therefore be more likely to purchase a shelter.

Other, more substantial objections to the economic approach have been made. Prospect theory offers the major alternative to the expected benefits (or utility) approach of economics (Kahneman and Tversky 1979). In prospect theory, decision-makers apply a two-stage process to risky choices, first editing or setting up a choice frame in which some risks are edited out of the problem, and then evaluating the included risks. Some low-probability risks are edited out and consequently ignored in decision-making; many experts contend that natural hazards risks are often ignored in this manner (Camerer and Kunreuther 1989; Meier 2006). Other approaches suggest that people apply a heuristic approach in which risks falling below some threshold probability are ignored, while risks perceived to exceed the threshold are addressed using the economic approach (Kunreuther 1996). In general, people have been characterized as myopic for ignoring (or excessively discounting) the risk associated with future events. With respect to tornado shelters, residents might focus on the out-of-pocket cost that must be incurred today (purchasing a tornado shelter might mean not buying a flat-screen HDTV or going on a vacation), while failing to consider the benefits of the shelter that may only become apparent in the distant future. Finally, inertia is an important factor in real-life decision-making: A resident who decides in principle to install a tornado shelter may procrastinate and never get around to calling a contractor (Zeckhauser and Samuelson 1986).

How accurately the economic model describes the decision-making of people with respect to a low-probability natural disaster is a subject of continuing debate (Camerer and Kunreuther 1989; Kunreuther and Pauly 2004). If people ignore tornado risk or other types of natural hazards, this may create a role for public policy. We will return to these issues with respect to the market for tornado shelters at the end of this chapter. In the meantime, we will use the economic model to offer some estimates of the cost-effectiveness of shelters.

5.3. The Cost-Effectiveness of Shelters by State

All of our calculations use several assumptions. We set the cost of an underground shelter at $3,000 including installation, based on prices of underground shelters available on the Internet in fall 2009. Safe rooms cost considerably more than this, typically $6,000 or more when built into a new home. In terms of safety, both underground shelters and safe rooms built to FEMA standards should protect residents from even the most powerful tornadoes, and thus we use the lower-cost underground shelters in our calculations. (We discuss the potential marginal benefits of a safe room relative to a shelter in Section 5.5.) We also make the heroic assumption that if shelters were installed in all homes (both permanent and mobile), all of the fatalities that occur in homes of this type would be avoided. This assumption is technically defensible because storm shelters are designed to withstand the strongest tornadoes. Thus, if residents were always able to reach their shelter before a tornado hit, fatalities could well be completely avoided. In practice, however, we know that people will not always make it into the shelter in time, for various reasons. Residents might be sleeping; they may not be watching TV or listening to the radio and therefore miss the warning; they may be unable to hear sirens; they may not see or hear the tornado in time to take cover; or the tornado might occur without any warning whatsoever. In addition, residents may choose to go outside and take pictures of the tornado, or attempt to find their pets and bring them to safety; moreover, injuries and even fatalities have occurred from falls while trying to reach shelter. By assuming shelters prevent all permanent- or mobile-home fatalities and injuries, we are calculating an upper bound on casualties avoided and have no available data to estimate how many fatalities might be expected from a failure to shelter in time or from falls. The figures we offer in this section can easily be adjusted for whatever percentage of current fatalities the reader thinks would not be avoided by the presence of a shelter. We use a 3% real discount rate in our calculations, which is approximately equal to the real interest rate available on mortgages as of fall 2009. Finally, we assume a useful life of 50 years for a shelter. In practice, the useful life of a shelter depends on its theoretical useful life and the expected life of the home itself—although homes and shelters can and do last more than 50 years, and even mobile homes have an estimated potential life in excess of 50 years, many homes are not actually used for this long and are abandoned or torn down. According to the 2000 census, the median year built for the census category one unit (detached housing units) was 1971, meaning that the median age of a home was only 29 years. In addition, only 22% of housing units were built before 1950. Thus, despite the potential for longer lives, it seems unlikely that homes and thus their storm shelters will have a useful life of more than 50 years, on average.

We provide state-by-state estimates of the cost per life saved for tornado shelters, with separate calculations for permanent and mobile homes. We make estimates using fatalities in a state and the probability of tornado damage. The calculations using state casualty totals estimate lives saved annually summed over all homes of the type in the state, and use a cost based on the total cost of installing shelters in all permanent or mobile homes in a state. We make one set of calculations using fatalities and injuries in a state between 1950 and 2007, and then infer the proportion of casualties we would expect in permanent and mobile homes without the widespread installation of shelters. We make a second set of calculations using the totals of permanent- and mobile-home fatalities in each state between 1996 and 2007. This allows us to avoid making inferences of how many fatalities occurred in each type of home in a state, but 12 years is a short time to try to estimate expected fatalities; the state totals could deviate substantially from true long-run rates over 12 years based on which states did and did not experience major tornado outbreaks during this time. We make calculations using the tornado probabilities in each state from Chapter 2, and fatalities and injuries per home struck by a tornado from Chapter 3. Fatalities actually observed in a state will depend on factors like the number of powerful tornadoes to strike populated areas; consequently, tornado probabilities might provide a better measure of the actual risk faced by residents. For example, a state with a high probability of damage might have avoided many powerful urban tornadoes and therefore have a relatively low fatality total.

The purpose of this analysis is to allow comparison of the value of tornado shelters across states, based on the frequency of tornadoes and incidence of casualties. The cost per life saved figures presented here can be compared with values of a statistical life estimated by economists based on market trade-offs to get a sense of where shelters offer cost-effective protection. Our estimates in no way constitute a recommendation for residents to purchase or not purchase a tornado shelter in any state. Even a resident of a state that has not experienced a tornado fatality since 1900 is not immune from tornado risk. Value in economics is subjective; the value of a tornado shelter to protect yourself and your family is equal to how much you place on safety. Different people place different values on safety, specifically protection from tornadoes—two neighbors might make different decisions regarding purchasing a shelter, and as economists we cannot say that either has made a mistake. No economist or government official can prescribe the value people "should" place on safety. Nonetheless, it is desirable that people make informed decisions regarding self-protection, and the comparative cost per fatality calculations offered in this chapter should be viewed in this light. If tornado shelters offer exceptional value, for instance, we might be concerned why the penetration rate of shelters is not higher.

Table 5.1 displays the cost per fatality avoided by the use of tornado shelters based on 1950–2007 fatality totals. The appendix to this chapter reports some illustrative calculations. The calculations in columns 1 and 2 apply the proportions of fatalities in permanent and mobile homes for the nation as a whole for the 1950–2007 period to each state. The first column in the table shows the cost per fatality avoided for permanent homes in millions of dollars by state, with state rankings in brackets. Mississippi and Arkansas, which ranked 1st and 2nd in fatalities per million residents, also have the lowest cost per fatality avoided in permanent homes at $39.0 million and $45.9 million, respectively. Alabama ($69 million), Kansas ($71 million), and Oklahoma ($78 million) round out the top five, and these are the only states with values below $100 million. The cost per fatality avoided escalates rapidly once outside of these five states. Only 13 states total have a cost per life saved of between $100 million and $200 million. The cost exceeds $1 billion per life saved in the 28th ranked state (Virginia), reaches nearly $36 billion in New Jersey, and is undefined for the seven states with no fatalities between 1950 and 2007. Econometric estimates of the value of a statistical life revealed in market trade-offs, as discussed in Chapter 4, fall in the range of $1 to $10 million, and so by this comparison tornado shelters in permanent homes fail to provide cost-effective protection in any state.

The second column in the table reports the cost per fatality avoided for mobile homes in millions of dollars, along with each state's rank. The lowest cost per fatality avoided occurs in Massachusetts at $3.3 million, although this figure appears to be an artifact of applying the national proportion of fatalities in mobile homes to each state, as we will discuss shortly. But even leaving Massachusetts aside, the cost per life saved for mobile homes is much lower than for permanent homes: Kansas ranks 2nd at $4.1 million, followed by Mississippi ($6.5 million), Arkansas ($6.7 million), and Oklahoma ($7.4 million). Seven states have a cost per life saved of less than $10 million, and 22 states have a cost below that of Mississippi's value for permanent homes. The difference in a state's ranking for permanent homes versus mobile homes depends on the proportion of mobile homes in a state: A state with a relatively small number of mobile homes will rank lower for mobile than for permanent homes.

That Massachusetts should have the lowest cost per fatality avoided for mobile homes seems implausible given the state's rank of 17th for permanent homes. Massachusetts's ranking results from applying the national percentage of fatalities nationwide in mobile homes, 43.2%, to the state. Mobile homes comprised less than 1% of Massachusetts housing units in 2000, and only 0.7% of state residents lived in mobile homes. Nationally, 7.6% of housing units were mobile homes and 6.9% of Americans lived in these homes in 2000, so we would not expect 43% of tornado fatalities in Massachusetts to occur in mobile homes. Yet the calculations reported in column 2 of Table 5.1 assume this, which projects to .76 mobile-home fatalities per year in Massachusetts and an excellent return on the cost of equipping the state's 24,000 mobile homes with shelters. The other states with significantly lower ranks for mobile than permanent homes also have a relatively small percentage of mobile homes: Connecticut (0.8%), Maryland (1.9%), Minnesota (4.5%), Nebraska (5.1%), and Wisconsin (4.3%). By contrast, mobile homes represent 20% of housing units in South Carolina, so we might expect more than 43% of fatalities in the state to occur in mobile homes.[4] The cost per fatality figures for mobile homes in Table 5.1 is most accurate for states with a housing stock similar to the nation as a whole.

We thus devised a formula to adjust the proportion of fatalities expected in a state in permanent and mobile homes based on the state's housing stock. Our adjustment assumes a linear relationship between the proportion of fatalities in each type of home and the proportion of each type of home in the state's housing stock. We can estimate a linear relationship because in addition to the observed proportion of fatalities for the national housing stock, we know that if there were no mobile (permanent) homes in a state, then no state fatalities would occur in mobile (permanent) homes. These two points allow us to fit a line. We calculate "nationwide" proportions of mobile and permanent homes in the housing stock by weighing the proportion of each type of home in a state by the state's share of tornado fatalities. The exact details of the formula are described in the appendix to this chapter.

Columns 3 and 4 of Table 5.1 report the cost per life saved for tornado shelters in permanent and mobile homes adjusted for the state housing stock. The differences in the cost per fatality avoided for permanent homes are relatively modest, typically within 5%, and exceeding 10% in only a few cases. In fact, 21 of the 41 states have the same ranking for permanent homes as with the national and state housing stock adjusted proportions of fatalities. The adjusted cost per life saved is slightly lower in the lowest-cost states; for instance, in Mississippi the cost per life saved falls from $39.0 million to $36.9 million. The adjustments produce more substantial changes for mobile homes, as intended. The cost per life saved in Massachusetts increases by

TABLE 5.1. Cost per Fatality of Tornado Shelters, by State

State	Historical Fatalities Permanent Homes	Historical Fatalities Mobile Homes	Housing Stock Adjusted Permanent Homes	Housing Stock Adjusted Mobile Homes
Alabama	$69.2 [3]	$11.3 [9]	$67.4 [4]	$7.4 [5]
Arizona	$7995.6 [39]	$1110.6 [41]	$9067.8 [39]	$847.9 [38]
Arkansas	$45.9 [2]	$6.7 [4]	$42.9 [2]	$4.7 [2]
Colorado	$5529.0 [37]	$347.9 [33]	$5740.8 [37]	$647.0 [37]
Connecticut	$4118.8 [34]	$40.3 [24]	$4505.1 [33]	$509.3 [34]
Delaware	$1835.1 [30]	$216.6 [32]	$2116.9 [30]	$204.8 [29]
Florida	$466.5 [25]	$62.1 [25]	$575.4 [25]	$56.3 [25]
Georgia	$242.9 [16]	$30.4 [22]	$243.8 [17]	$26.7 [17]
Illinois	$283.8 [19]	$10.3 [8]	$315.7 [20]	$34.0 [20]
Indiana	$157.2 [8]	$9.7 [7]	$142.3 [7]	$15.5 [7]
Iowa	$275.8 [18]	$12.8 [13]	$240.3 [16]	$25.8 [16]
Kansas	$70.7 [4]	$4.1 [2]	$62.9 [3]	$6.8 [3]
Kentucky	$195.0 [13]	$27.7 [21]	$190.3 [12]	$20.8 [13]
Louisiana	$153.1 [9]	$20.7 [19]	$153.9 [9]	$16.8 [10]
Maine	$7078.3 [38]	$845.8 [39]	$6767.6 [38]	$910.5 [39]
Maryland	$3148.2 [33]	$74.7 [26]	$3965.7 [32]	$408.7 [32]
Massachusetts	$263.5[17]	$3.3 [1]	$324.0[21]	$37.9 [22]
Michigan	$235.2 [15]	$14.3 [16]	$214.8 [14]	$23.1 [16]
Minnesota	$288.5 [20]	$12.0 [11]	$274.4 [18]	$27.9 [18]
Mississippi	$39.0 [1]	$6.5 [3]	$36.9 [1]	$4.2 [1]
Missouri	$161.9 [10]	$12.4 [12]	$151.7 [10]	$16.0 [9]
Montana	$2546.3 [31]	$393.0 [35]	$2449.8 [31]	$290.3 [31]
Nebraska	$190.9 [12]	$8.2 [6]	$171.0[11]	$16.7 [11]
New Jersey	$35633.8 [41]	$439.0 [36]	$42355.8 [41]	$4563.4 [41]
New Mexico	$1779.9 [29]	$374.8 [34]	$1881.9 [29]	$229.6 [30]
New York	$2944.4 [32]	$123.9 [31]	$4556.4 [34]	$484.2 [33]
North Carolina	$445.9 [24]	$77.4 [27]	$446.6 [24]	$49.9 [23]
North Dakota	$141.4 [7]	$13.5 [15]	$146.9 [8]	$15.9 [8]
Ohio	$352.1 [22]	$16.0 [17]	$337.0 [22]	$36.8 [21]
Oklahoma	$78.3 [5]	$7.4 [5]	$70.7 [5]	$7.3 [4]
Pennsylvania	$692.4 [27]	$39.8 [23]	$798.2 [27]	$85.3 [27]
South Carolina	$381.5 [23]	$86.2 [28]	$399.8 [23]	$53.0 [24]
South Dakota	$232.9 [14]	$26.1 [20]	$222.9 [15]	$24.2 [15]
Tennessee	$119.7 [6]	$13.3 [14]	$114.6 [6]	$12.7 [6]
Texas	$188.8 [11]	$17.2 [18]	$191.9 [13]	$20.2 [12]
Utah	$10255.7 [40]	$481.5 [38]	$9768.3 [40]	$994.4 [40]
Virginia	$1326.5 [28]	$90.1 [29]	$1371.6 [28]	$149.0 [28]
Washington	$5038.3 [35]	$474.7 [37]	$5209.6 [36]	$590.7 [35]
West Virginia	$5466.3 [36]	$928.3 [40]	$5098.1 [35]	$602.3 [36]
Wisconsin	$293.7 [21]	$12.0 [10]	$286.9 [19]	$29.0 [19]
Wyoming	$676.8 [26]	$112.6 [30]	$672.3 [26]	$74.8 [26]

Amounts in millions of dollars per life saved for tornado shelters. State rankings in [ ]. Values are undefined and omitted from the rankings for California, Idaho, Nevada, New Hampshire, Oregon, Rhode Island, and Vermont. These states experienced no fatalities between 1950 and 2007.

more than an order of magnitude, from $3.3 million to almost $38 million, and the state now ranks 22nd, in line with its adjusted ranking for permanent homes (21st). The cost per life saved similarly increases by an order of magnitude due to a small proportion of mobile homes in several other states, including Connecticut and New Jersey, even though New Jersey already ranked 36th for mobile homes with the unadjusted figures. The cost figure declines with our adjustment for several states with a large fraction of mobile homes, including a 33% reduction in Mississippi, which now has the lowest cost per life saved at $4.2 million, as well as Arkansas and Alabama. Tornado shelters still provide good value for mobile-home residents in Kansas at $6.8 million, but this figure is over 60% higher when adjusted for the state's relatively small proportion of mobile homes. The cost per fatality avoided is also under $10 million in Oklahoma. Twenty-seven states have a cost per life saved below $100 million for mobile homes, so although shelters provide more cost-effective protection for mobile homes than permanent homes, the cost per life saved is nevertheless well above the range of values observed in market trade-offs in many states.

The calculations in Table 5.1 consider the cost-effectiveness of shelters only in terms of fatalities avoided, and the reader might wonder how our assessment would change if injuries were included. We can calculate a cost per casualty avoided by combining or "adding" fatalities and injuries. Since a fatality is a much worse outcome than an injury, however, simply adding the two together would not be appropriate. Injuries must be discounted, and the question then is how much to discount injuries relative to fatalities. The relative values of statistical lives and injuries provide a reasonable way to weight injuries versus fatalities. The choice of a weight depends on the distribution of the severity of tornado injuries. Available epidemiological studies suggest that injuries are not severe. Brown et al. (2002), for example, found that 76% of injuries in the May 3, 1999, Oklahoma tornado outbreak did not require hospitalization, and hospital stays when required averaged 7 days. Carter et al. (1989) found that 83% of injuries in the May 31, 1985, Ontario, Canada, tornado outbreak were minor, with serious injuries requiring an average 12.5-day hospital stay. Consequently, we will equate injuries to fatalities at the rate of 100 to 1, as we have in previous work on shelters (Merrell, Simmons, and Sutter 2005; Simmons and Sutter 2006), so an average of 50 injuries per year in a state would be equivalent to .5 deaths per year. We also lack data on the breakdown of tornado injuries by location, and so we must infer a distribution of injuries in permanent and mobile homes. We will apply the proportions of fatalities in mobile and permanent homes to injuries, although it is likely that a higher proportion of injuries than fatalities occur in permanent homes.[5] We will discuss in the next section how different distributions of injuries affect the cost per casualty avoided.

Table 5.2 reports the cost per casualty avoided calculations across states, using the state housing stock adjusted proportions of fatalities and injuries in mobile and permanent homes, as in the final two columns of Table 5.1. All 48 contiguous states have experienced injuries, so we can calculate a cost per casualty avoided for all states. The cost per casualty avoided is quite similar to the cost per fatality avoided, indicating that the inclusion of injuries, at least at the 100 to 1 ratio, has little impact on the cost-effectiveness of shelters. Nationally there were 81,000 injuries and almost 4,900 fatalities between 1950 and 2007, so the injuries are the equivalent of about 800 fatalities, or one-sixth the number of fatalities. Adding injuries to fatalities reduces the cost per casualty by about 15% relative to the cost per fatality. The cost per fatality and casualty avoided for both types of homes are highly correlated (+.99 in each case), and the small differences in cost-effectiveness are a function of differing numbers of injuries per fatality across states. Injuries per fatality varied from 9.3 in Missouri to 175 in Connecticut. Higher ratios of injuries per fatality tend to occur in states with relatively few fatalities; for instance, 10 states have experienced 30 or more injuries per fatality, but none of these states had more than 6 fatalities, and the correlation between injuries per fatality and total fatalities is −0.40. The states where inclusion of injuries had the greatest effect on the cost per casualty already had very high costs per life saved. The cost per casualty avoided for shelters in Connecticut is about one-third of the cost per fatality, but the cost per casualty avoided is still $1.6 billion. The cost per casualty avoided is $32 million in permanent homes in Mississippi, and less than $6.4 million for mobile homes in five vulnerable states.

The estimates in Tables 5.1 and 5.2 had to infer the proportion of tornado fatalities in permanent and mobile homes in each state based on the location of fatalities across the nation. An alternative would be to use actual totals of permanent- and mobile-home fatalities in each state. Table 5.3 presents cost per fatality avoided calculations for shelters in permanent and mobile homes based on observed fatalities locations for the period over which we have detailed information on all fatalities, 1996–2007. Since we do not have detailed information on the location of injuries, Table 5.3 reports only cost per fatality figures. Twelve years represents a very short period of time in which to estimate casualties, and thus the state totals will tend to deviate more from their true, unobserved long run totals. States that experienced

TABLE 5.2. Cost per Casualty Avoided, based on Adjusted Proportions, 1950–2007

State	Permanent Homes	Mobile Homes
Alabama	$58.6 [4]	$6.4 [5]
Arizona	$6196.6 [40]	$579.4 [39]
Arkansas	$37.6 [2]	$4.2 [2]
California	$181,310 [46]	$19,776.1 [46]
Colorado	$3,865.9 [36]	$435.7 [37]
Connecticut	$1,638.2 [31]	$185.2 [31]
Delaware	$1,550.8 [30]	$150.1 [29]
Florida	$477.5 [25]	$46.7 [25]
Georgia	$201.0 [17]	$22.0 [17]
Idaho	$58,971.8 [45]	$6,629.7 [45]
Illinois	$263.0 [20]	$28.3 [20]
Indiana	$121.6 [7]	$13.2 [7]
Iowa	$184.8 [15]	$19.9 [15]
Kansas	$56.4 [3]	$6.1 [3]
Kentucky	$154.1 [12]	$16.8 [12]
Louisiana	$131.3 [8]	$14.3 [10]
Maine	$5,687.1 [39]	$765.1 [40]
Maryland	$2,751.2 [33]	$283.5 [33]
Massachusetts	$285.9 [22]	$33.4 [22]
Michigan	$188.7 [16]	$20.3 [16]
Minnesota	$229.1 [18]	$23.3 [18]
Mississippi	$32.2 [1]	$3.7 [1]
Missouri	$132.2 [10]	$14.0[11]
Montana	· $2197.1 [32]	$260.3 [32]
Nebraska	$140.7 [11]	$14.0 [8]
Nevada	$529,162 [47]	$58,337.9 [47]
New Hampshire	$22,722.0 [41]	$2,744.3 [41]
New Jersey	$25,670.2 [43]	$2,765.7 [42]
New Mexico	$1,436.6 [29]	$175.3 [30]
New York	$3,975.3 [37]	$422.4 [36]
North Carolina	$365.2 [24]	$40.8 [23]
North Dakota	$129.1 [9]	$14.0 [9]
Ohio	$272.0 [21]	$29.6 [21]
Oklahoma	$61.2 [5]	$6.3 [4]
Oregon	$612,069 [48]	$70,425.7 [48]
Pennsylvania	$694.8 [27]	$74.2 [27]
Rhode Island	$24,345.5 [42]	$2780.9 [43]
South Carolina	$323.8 [23]	$42.9 [24]
South Dakota	$177.4 [14]	$19.3 [14]
Tennessee	$100.9 [6]	$11.1 [6]
Texas	$166.6 [13]	$17.5 [13]
Utah	$5,061.3 [38]	$515.2 [38]
Vermont	$32,262.6 [44]	$4133.2 [44]
Virginia	$1,143.0 [28]	$124. [29]
Washington	$3,461.5 [35]	$392.5 [34]
West Virginia	$3,365.1 [34]	$397.6 [35]
Wisconsin	$247.0 [19]	$25.0 [19]
Wyoming	$542.2 [26]	$60.3 [26]

Amounts in millions of dollar per life saved by tornado shelters. State rankings in [ ].

TABLE 5.3. Cost per Fatality for Shelters, 1996–2007

State	Permanent Homes	Mobile Homes
Alabama	$42.4 [1]	$9.4 [5]
Arkansas	$62.2 [6]	$7.7 [2]
Colorado	$1424.4 [22]	$124.4 [24]
Florida	$600.9 [15]	$16.8 [8]
Georgia	$891.6 [17]	$9.1 [4]
Illinois	$3692.3 [25]	$26.6 [12]
Indiana	Undefined	$7.9 [3]
Iowa	$586.3 [14]	$75.8 [19]
Kansas	$61.1 [5]	$14.1 [7]
Kentucky	$728.4 [16]	$35.9 [14]
Louisiana	$1478.6 [23]	$21.4 [11]
Maryland	$473.1 [13]	Undefined
Michigan	Undefined	$77.5 [20]
Minnesota	$349.4 [11]	Undefined
Mississippi	$98.5 [8]	$57.2 [18]
Missouri	$78.0 [7]	$18.7 [9]
Nebraska	$331.9 [10]	Undefined
New Mexico	Undefined	$83.8 [22]
North Carolina	$2785.9 [24]	$51.7 [17]
North Dakota	Undefined	$29.0 [13]
Ohio	$1043.2 [18]	$87.8 [23]
Oklahoma	$49.5 [3]	$13.4 [6]
Pennsylvania	$1233.8 [20]	Undefined
South Carolina	$1326.8 [21]	$82.3 [21]
South Dakota	$45.0 [2]	$42.0 [15]
Tennessee	$57.8 [4]	$7.6 [1]
Texas	$191.7 [9]	$43.4 [16]
Virginia	$1153.4 [19]	Undefined
Wisconsin	$374.6 [12]	Undefined
Wyoming	Undefined	$20.1 [10]

Cost per fatality avoided for tornado shelters, based on fatalities by location over the period 1996–2007, in millions of dollars.

one or more particularly deadly tornadoes over these years might be well above their long run values, while many states did not experience a fatality in these locations within the time frame even though their long-run risk is not zero. Table 5.3 reports the cost-effectiveness figures only for states with fatalities in either permanent or mobile homes, and readers should understand that these figures contain noise. Twenty-five states have a cost per fatality for permanent homes; the lowest is in Alabama ($42 million per life saved), followed by South Dakota ($45 million), Oklahoma ($50 million), Tennessee ($58 million), and Kansas ($61 million). The cost per life saved for permanent homes is also less than $100 million in Arkansas, Missouri, and Mississippi. Tennessee has the lowest cost per life saved for mobile homes at $7.6 million, followed by Arkansas ($7.7 million), Indiana ($7.9 million), Georgia ($9.1 million), and Alabama ($9.4 million). Although the rank of states does differ from Table 5.2, the states with the lowest cost per life saved over the 12-year period also tend to be tornado prone.

One interesting note to the dollar figures in Table 5.3 is that the lowest cost per life saved figures are very similar to those for the 1950–2007 period, although they occur in different states. In fact, the lowest cost per life saved for permanent homes in Table 5.3, $42 million in Alabama, exceeds Mississippi's value of $37 million in column 3 of Table 5.1, and Tennessee's $7.6 million for mobile homes exceeds Mississippi's $4.2 million in column 4 of Table 5.1. This is surprising, since over a period of only 12 years we might expect that at least several states would have experienced a couple of deadly tornadoes to produce a very low cost per fatality avoided. Yet this has not occurred. Alabama and Tennessee have the highest fatality totals over the 1996–2007 period, and although shelters offer better value in these states than when based on fatalities over the 1950–2007 period, the costs for permanent homes still exceed values of statistical lives observed in market trade-offs. Florida and Georgia were identified as outliers for mobile-home fatalities in Section 3.6 and yet do not have the lowest costs per fatality avoided for shelters in mobile homes in Table 5.3; in fact, Florida ranks only 8th at $16.8 million. Fatalities in permanent homes appear to be too infrequent to result in a low cost per life saved figure, even when based on a relatively short period of time.

We can calculate a cost per life saved more directly using the probability of a tornado in a state and the probability of a fatality or casualty. This approach allows the calculation of a site-specific value of a tornado shelter. The probability of a tornado fatality in a particular residence over a year is the product of three probabilities: the probability that a tornado will strike the residence in the year; the probability that residents will be home when the tornado strikes; and the probability that the residents at home when a tornado strikes are killed. The probability of a tornado was estimated in Chapter 2 and based on damage paths since 1950. A lack of data precludes separate estimation of the second and third probabilities, but their product can be approximated using fatalities per residence struck by a tornado, as estimated in Chapter 3 (see Table 3.11). To separately estimate the cost per life saved for permanent and mobile homes, we take the fatalities per home struck from Table 3.11 (.0217 for mobile homes and .00246 for permanent homes) and apply these fatality rates in all states. Fatalities per residence damaged or destroyed in a tornado multiplied by the annual tornado probability for each state yields expected fatalities per year for a residence of each type in each state. We also use the injuries per residence struck for permanent and mobile homes reported in Table 3.11 to calculate a cost per casualty avoided. We can calculate the cost per life saved for tornado shelters straightforwardly at this point, assuming that a shelter prevents all of these fatalities. For an example of this calculation, see the appendix to this chapter. Note that more refined, location-specific estimates of tornado probabilities can be substituted into this formula.

Table 5.4 reports the cost-effectiveness of shelters for permanent and mobile homes based on the probability of tornado damage. A cost per life saved can be calculated for each state with this method since all states experienced tornadoes. The use of probabilities of tornado damage and nationwide casualties per residence struck adjusts for states that experienced a particularly high or low number of fatalities relative to the number of tornadoes. If a state were fortunate enough to avoid loss of life in tornadoes since 1950, historical fatalities would be low and the cost per life saved for shelters artificially high.[6] The lowest cost per life saved in permanent homes by this method occurs in Mississippi at $111 million, followed by Arkansas ($113 million), Oklahoma ($119), Kansas ($140 million), and Iowa ($141 million). In particular, Iowa moves from 14th with a cost of $131 million per life saved based on fatalities to 5th place based on tornado probability. This indicates that Iowa's fatality rate has been relatively low for the state's underlying tornado rate. Eighteen states have a cost per life saved in excess of $1 billion, with Nevada leading the way at $98 billion per life saved. Among states with no fatalities between 1950 and 2007, most have extremely high cost per life saved using tornado probabilities. The exception is Rhode Island, which ranks 24th with a cost per life saved of $467 million. Rhode Island actually ranks in the middle of states based on tornado climatology, but has not experienced a fatality over the period. Some other states whose ranks have changed noticeably include South Carolina, New Jersey, North Carolina, and Wisconsin, with lower costs per life saved in Table 5.4, while Massachusetts and North Dakota have much higher costs per life saved. For mobile homes, the cost per life saved in the five states with the highest tornado probabilities is $12.6 million in Mississippi, $12.8 million in Arkansas, $13.5 million in Oklahoma, $15.9 million in Kansas, and $16.0 million in Iowa. The same probabilities of injuries in permanent homes and mobile homes are applied in all states, so the cost per casualty avoided falls by the same percentage in each case. The permanent-home cost falls by about 20% because the probability of injury is about 10 times greater than the probability of fatality, while the cost for mobile homes is only about 2% lower when including injuries.

TABLE 5.4. Cost-Effectiveness of Shelters on Tornado Probabilities, 1950–2006

State	Cost per Fatality Avoided		Cost per Casualty Avoided	
	Permanent Homes	Mobile Homes	Permanent Homes	Mobile Homes
Alabama	$176.2 [7]	$20.0 [7]	$140.5 [7]	$19.4 [7]
Arizona	$28156.8 [45]	$3191.4 [45]	$22452.6 [45]	$3099.9 [45]
Arkansas	$112.5 [2]	$12.8 [2]	$89.7 [2]	$12.4 [2]
California	$28035.7 [44]	$3177.6 [44]	$22356.1 [44]	$3086.5 [44]
Colorado	$2554.9 [34]	$289.6 [34]	$2037.3 [34]	$281.3 [34]
Connecticut	$750.5 [30]	$85.1 [30]	$598.4 [30]	$82.6 [30]
Delaware	$622.8 [27]	$70.6 [27]	$496.7 [27]	$68.6 [27]
Florida	$691.3 [29]	$78.4 [29]	$551.2 [29]	$76.1 [29]
Georgia	$214.9 [12]	$24.4 [12]	$171.3 [12]	$23.7 [12]
Idaho	$22915.1 [43]	$2597.3 [43]	$22356.1 [43]	$2522.8 [43]
Illinois	$192.9 [8]	$21.9 [8]	$153.9 [8]	$21.2 [8]
Indiana	$154.1 [6]	$17.5 [6]	$122.9 [6]	$17.0 [6]
Iowa	$141.3 [5]	$16.0 [5]	$112.6 [5]	$15.6 [5]
Kansas	$140.3 [4]	$15.9 [4]	$111.8 [4]	$15.4 [4]
Kentucky	$383.2 [22]	$43.4 [22]	$305.5 [22]	$42.2 [22]
Louisiana	$256.1 [13]	$29.0 [13]	$204.2 [13]	$28.2 [13]
Maine	$11333.1 [41]	$1284.5 [41]	$9037.1 [41]	$1247.7 [41]
Maryland	$627.9 [28]	$71.2 [28]	$500.7 [28]	$69.1 [28]
Massachusetts	$419.6 [23]	$47.6 [23]	$334.6 [23]	$46.2 [23]
Michigan	$282.2 [15]	$32.0 [15]	$225.1 [15]	$31.1 [15]
Minnesota	$357.2 [20]	$40.5 [20]	$284.9 [20]	$39.3 [20]
Mississippi	$111.1 [1]	$12.6 [1]	$88.6 [1]	$12.2 [1]
Missouri	$262.0 [14]	$29.7 [14]	$208.9 [14]	$28.8 [14]
Montana	$7215.6 [39]	$817.8 [39]	$5753.9 [39]	$794.4 [39]
Nebraska	$199.1 [11]	$22.6 [11]	$158.8 [11]	$21.9 [11]
Nevada	$98483.8 [48]	$11162.4 [48]	$78532.4 [48]	$10842.4 [48]
New Hampshire	$3155.8 [36]	$357.7 [36]	$2516.5 [36]	$347.4 [36]
New Jersey	$612.7 [26]	$69.4 [26]	$488.6 [26]	$67.5 [26]
New Mexico	$10918.8 [40]	$1237.6 [40]	$8706.8 [40]	$1202.1 [40]
New York	$1138.5 [32]	$129.0 [32]	$907.8 [32]	$125.3 [32]
North Carolina	$300.7 [17]	$34.1 [17]	$239.8 [17]	$33.1 [17]
North Dakota	$1389.2 [33]	$157.5 [33]	$1107.7 [33]	$152.9 [33]
Ohio	$286.9 [16]	$32.5 [16]	$228.8 [16]	$31.6 [16]
Oklahoma	$119.1 [3]	$13.5 [3]	$95.0 [3]	$13.1 [3]
Oregon	$28455.1 [46]	$3225.2 [46]	$22690.5 [46]	$3132.7 [46]
Pennsylvania	$348.3 [19]	$39.5 [19]	$277.7 [19]	$38.3 [19]
Rhode Island	$467.0 [24]	$52.9 [24]	$372.4 [24]	$51.4 [24]
South Carolina	$328.5 [18]	$37.2 [18]	$261.9 [18]	$36.2 [18]
South Dakota	$532.1 [25]	$60.3 [25]	$424.3 [25]	$58.6 [25]
Tennessee	$195.0 [9]	$22.1 [9]	$155.5 [9]	$21.5 [9]
Texas	$360.5 [21]	$40.9 [21]	$287.4 [21]	$39.7 [21]
Utah	$17701.6 [42]	$2006.3 [42]	$14115.5 [42]	$1948.8 [42]
Vermont	$3365.7 [37]	$381.5 [37]	$2863.9 [37]	$370.5 [37]
Virginia	$1042.9 [31]	$118.2 [31]	$831.6 [31]	$114.8 [31]
Washington	$31162.8 [47]	$3532.1 [47]	$24849.7 [47]	$3430.8 [47]
West Virginia	$6748.6 [38]	$764.9 [38]	$5381.4 [38]	$743.0 [38]
Wisconsin	$196.0 [10]	$22.2 [10]	$156.3 [10]	$21.6 [10]
Wyoming	$2958.0 [35]	$335.3 [35]	$2358.8 [35]	$325.7 [35]

The costs per life saved based on probabilities are substantially based on fatalities. This may be because the years over which we estimated fatalities per home, 2002–2004, were relatively low-fatality years and occurred late in the 1950–2007 period. Fatalities per home in these years may not reflect casualties in the 1950s. We will return to the impact of the downward time trend in fatalities on our calculations in the next section.

5.4. Extensions of Our Analysis

We have used several methods to estimate the value of tornado shelters, and in this section we extend our analysis to consider several additional factors. Specifically, we will consider the potential for our estimates to change over time, the effect of the downward trend in lethality of tornadoes on tornado shelters, and how the value of peace of mind affects our assessment of shelters.

Even over a period of nearly 60 years, both tornado casualties and tornado damage areas provide imperfect evidence about the true, unobserved tornado probability or expected annual tornado fatalities in a state as discussed in Chapter 2. The cost per life saved for shelters could be updated and refined each year based on the number of tornadoes and fatalities in the state in that particular year. Residents should be concerned about the potential variations in the cost-effectiveness of shelters based on this new information. The potential shelter buyer will want to know if the estimates reported here are likely to change significantly with a few more years' data, because the purchase of a shelter is irreversible and the cost of retrofitting a house with a safe room (instead of designing the room into new construction) is considerable. In cases where such an irreversible decision must be made, delaying the decision to wait for improved information can yield significant value (Arrow and Lind 1970; Arrow and Fisher 1974). On the other hand, the buyer might decide not to purchase a shelter based on the current cost per fatality avoided, but then regret this decision if a major tornado causes a significant downward revision of the cost. One storm can matter significantly even over 50 years or more of data, as Massachusetts illustrates. Ninety of the 102 tornado fatalities in Massachusetts occurred in one storm, the 1953 Worcester F4 tornado. Suppose that the Worcester tornado had instead occurred in 2007, and that the state had experienced only 12 fatalities in the 57 years from 1950 through 2007. Fatalities per year in Massachusetts would increase from 0.21 to 1.76 through 2007, and the cost per life saved

TABLE 5.5. The Change in Cost per Fatality Avoided with Shelters, 1999–2007

State	Percentage Reduction in Cost per Life Saved, 1950–1999 to 1950–2007	Cost per Life Saved, for 1950–1999
Maryland	−66.9%	$9499
Wyoming	−42.0%	$1167
Colorado	−42.0%	$9533
New Mexico	−30.4%	$2557
Tennessee	−12.5%	$137
Missouri	−11.0%	$182
Georgia	−9.9%	$271
Florida	−5.8%	$495

The cost figure is for permanent homes based on historical fatalities over 1950–1999, in millions of dollars.

in permanent homes would decline from $2.7 billion to the $324 million reported in Table 5.1, an 88% decline.

Although extreme, this case illustrates that estimates of cost-effectiveness have the potential to change substantially in a short time. We explore the change in the cost per life saved more typically observed in practice based on tornado casualties. For casualties, we calculated the cost per life saved in each state based on fatalities over the 1950–1999 period, and the percentage change in this value compared to the 1950–2007 values in Table 5.1. Thus we construct an experiment using 8 years of extra data on fatalities to update estimates based on 50 years of data. If no fatalities occurred in a state in these 8 years, the cost per fatality avoided increases by 16%. Table 5.5 reports the cost per fatality avoided through 1999 and the percentage change in this figure with the 8 extra years of data for the states with a reduction over the period. Several states experienced a substantial decline in the cost per fatality avoided, including a 42% decline in Colorado and Wyoming and a 67% decline in Maryland. In addition, the cost per fatality avoided became defined in New Jersey after the state had no fatalities during the 1950–1999 period. Maryland exhibits the largest decline in cost, with 5 fatalities between 2000 and 2007 after only 2 fatalities between 1950 and 1999. The cost per fatality avoided through 1999 in Maryland was $9.5 billion, and so the 67% reduction still leaves the state with a cost in excess of $3.1 billion. The cost per life saved through 1999 is also high in the other states in Table 5.5, in excess of $1.1 billion in the four states with a 30% or larger reduction, and between $140 million and $500 million for the other states experiencing reductions. Thus, at least based on the number of fatalities observed in recent years, the cost per life saved calculations for shelters are reasonably stable. A large percentage reduction is only possible when the initial cost per life saved is extremely high, and remains beyond the observed values of a statistical life after being updated.

Another perspective on the potential for new information to substantially revise estimates is provided by the hypothetical case of a tornado equivalent to the May 3, 1999, Oklahoma City F5 tornado occurring in the state during 2008. The May 3 tornado killed 36 persons, the highest total for a tornado since 1979. We recalculated the cost per life saved using 1950–2007 fatalities and assumed 36 fatalities in 2008. Needless to say, the cost per fatality avoided falls dramatically in many states, but again the biggest percentage change occurs in states with few fatalities through 2007; and for the seven states with no fatalities since 1950, the percentage reduction is infinite. Again, though, the states with a large percentage reduction in the cost per life saved also had a very high cost per life saved even after the hypothetical 36 fatalities in 2008. The most relevant cases for examination would be the percentage reduction in cost per fatality avoided in states with a cost under $100 million after the hypothetical killer tornado. The percentage reductions in cost in these states were −8.1% in Arkansas, −12.8% in Kansas, −6.5% in Mississippi, −60.3% in North Dakota, and −10.4% in Oklahoma. The notable outlier here is North Dakota; a 36-fatality tornado would more than double the state's total number of fatalities. The other states had fatality totals over 200, so the percentage change in cost-effectiveness would be small for most states with a high enough fatality rate to have a cost per life saved under $100 million.

The figures for fatalities and injuries per year in a state used in the calculations in Tables 5.1 and 5.2 are the averages over the 58-year period from 1950 to 2007. However, as we saw in the regression analysis in Chapter 3, fatalities have trended down over this time. The point estimate of a linear time trend indicated a 29% reduction in fatalities by 2007 compared with 1950. By applying average fatalities over these years, we are allowing fatalities from tornadoes in the 1950s to inflate the benefits from shelters today, even though these tornadoes would kill fewer people if they occurred today. We can see evidence of the downward trend in the lethality of tornadoes when we consider the calculations in Table 5.3 using fatalities by location over the 1996–2007 period. Twelve years is a short period of time to assess tornado risk, and thus we would expect to see some states have very low costs per life saved due to one particularly deadly tornado outbreak during these years. And yet, the lowest cost per life saved in Table 5.3 exceeds the lowest costs in Table 5.1, for both permanent and mobile homes. In addition, the cost per life saved based on tornado probabilities in Table 5.4 are substantially higher; these estimates are based on fatalities per home struck in the years 2002–2004, and might be further evidence of lower current casualty rates.

We can adjust the cost per life saved calculations in a straightforward way to reflect the downward trend in tornado lethality. Fatalities at the end of the 1950–2007 period were about 30% lower than at the beginning. The fatalities per year used in our calculations reflect the average over the period, and given the 30% reduction over the period, fatalities in 2007 would be expected to be about 17% lower than the average over the period. A 17% reduction in fatalities per year in a state will increase the cost per life saved for both permanent and mobile homes by about 20%. This adjustment would increase the cost per life saved for shelters in Mississippi based on the adjustments for state housing stock to about $45 million in permanent homes and over $5 million in mobile homes. The adjusted cost per life saved for mobile homes in Kansas, which ranks 3rd, is now over $8 million, and so even in tornado-prone states, shelters are on the margin of market-revealed values of a statistical life in mobile homes. We have not tried to refine any state-specific time trends in fatalities, so it is possible that some states have experienced reductions in fatalities more or less than 30% since 1950.

Information is not available on locations where tornado injuries occur. As a consequence, our attribution of injuries to permanent or mobile homes for calculation of the cost per fatality avoided could easily be over- or underestimated, and because of this uncertainty we investigate the sensitivity of our calculations to a variation in injury locations. Specifically, we consider how the cost per casualty would vary in each state and for each type of home, as the percentage of all injuries occurring in homes of this type ranges from 0 (the adjusted cost per life saved figures in Table 5.1) to 100. In some cases, the percentage reduction in the cost per life saved is substantial, up to 86% and 98% reductions in permanent and mobile homes, respectively, in Connecticut. The percentage reduction is larger in states with low fatality rates, and a small proportion of mobile-home housing units also increases the percentage reduction. The reduction in cost for permanent homes is typically 30% to 40% for states with a cost per life saved in Table 5.1 of less than $300 million, and a 20%–30% reduction for mobile homes is also typically observed in high-fatality states. Figures 5.1 and 5.2 show the cost-effectiveness of shelters in Mississippi. For permanent homes, the cost per casualty avoided falls from $37 million with no injuries included to $26 million if all Mississippi injuries occurred in permanent homes, while the cost per casualty avoided for mobile homes falls from $4.2 million to $3.5 million. Obviously

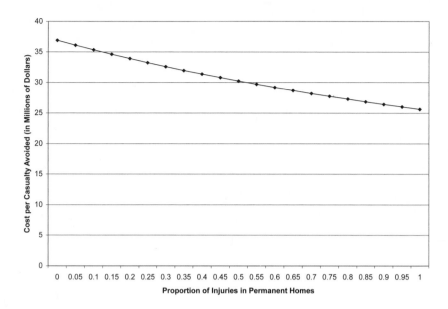

FIGURE 5.1. Injuries and the cost-effectiveness of shelters (permanent homes). The figure shows the cost per casualty avoided for permanent homes in Mississippi in millions of dollars as a function of the proportion of state injuries in permanent homes.

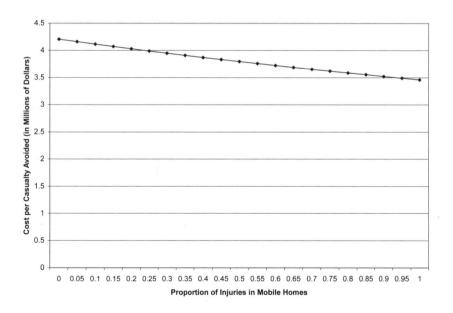

FIGURE 5.2. Injuries and the cost-effectiveness of shelters (mobile homes). The figure shows the cost per casualty avoided for mobile homes in Mississippi in millions of dollars as a function of the proportion of state injuries in mobile homes.

the percentage of injuries in these two locations cannot sum to over 100, so the cost per casualty avoided in Mississippi cannot simultaneously be $26 million for permanent homes and $3.5 million for mobile homes. We see that the apportionment of injuries will not change our conclusions about shelters: Shelters are not cost-effective (relative to market-revealed values of a statistical life) in permanent homes, but are cost-effective for mobile homes in the most tornado-prone states.

Our calculations value shelters only through the fatalities and injuries they prevent. Still, residents surely benefit from a shelter even if it is never struck by a tornado: The benefit is the reduced anxiety experienced by residents if they shelter in a safe room during a tornado warning. Heightened stress was reported among Oklahomans during severe weather threats after the May 3, 1999, tornadoes, and shelters would reduce such stress. A peace-of-mind benefit would reduce the cost per life saved required for shelters to be cost-effective, and since the peace-of-mind benefit occurs whenever residents must shelter instead of only when the home is struck by a tornado, the very small annual probability of a tornado strike does not cause the expected peace-of-mind benefit to necessarily diminish dramatically.

We cannot place a dollar value directly on the peace-of-mind benefit of a tornado shelter, because to date survey research has not explored residents' anxiety per warning or per minute of a warning. We can, however, use NOAA tornado-warning records to help approximate the potential peace-of-mind benefit. We tabulated warning totals for U.S. counties in Chapter 4; Table 4.5 reported the most warned counties. We use these county warning totals to estimate the frequency of tornado warnings. Table 5.6 reports the frequencies for the five states with the lowest cost per life saved for shelters in permanent homes: Alabama, Arkansas, Kansas, Mississippi, and Oklahoma. The table reports two rates for each state, one averaged for all counties in the state, and the second for the most warned county in the state, which we offer as an upper bound on warning frequency. Warnings per year are tabulated over two different time periods, 1986–2004 and 1996–2004. Although a longer time period usually allows a better estimate of frequencies, NWS tornado warnings per year increased significantly in the mid 1990s with the implementation of Doppler weather radars; for this reason, averages over the 1996–2004 period may provide a better estimate of warning frequency in the future. On the other hand, Storm Based Warnings for tornadoes implemented by the NWS in October 2007 reduce the warned area and thus the frequency of warning for any given home. The advent of SBWs suggests that warning frequency in the future will be at or below the 1986–2004 rate. Over

TABLE 5.6. Warnings per Year for Counties in Tornado-Prone States

State	Warnings/Year, 1986–2004	Warnings/Year, 1996–2004
Alabama, All Counties	1.95	3.06
Alabama, Baldwin	4.95	6.44
Arkansas, All Counties	1.01	1.64
Arkansas, White	2.42	4.22
Kansas, All Counties	1.21	1.45
Kansas, Sumner	3.32	4.44
Mississippi, All Counties	1.81	2.50
Mississippi, Harrison	4.32	5.89
Oklahoma, All Counties	1.65	2.28
Oklahoma, Osage	3.89	5.78

the longer period, counties in each of these states averaged between 1 and 2 warnings per year, led by Alabama at 1.95 and Mississippi at 1.81 warnings per county per year. Since 1996, Alabama averaged 3 warnings per county per year. The most warned county in each state was warned more than twice as often as the average for all counties in the state. Over a period of two decades the most warned counties are particularly likely to have recorded warning totals greater than their long-run rates. In any event, a household in any of these tornado-prone states can probably expect to be warned once per year in the future, and as discussed in Chapter 4, the typical warning is in effect for about 40 minutes.

So the peace-of-mind benefit arises from the value of reduced anxiety in about one warning per year, for perhaps 40 minutes. We have no reference to value this time, so we approach this problem in two different ways. First, we can calculate the present value of peace-of-mind benefits over the life of the shelter assuming that the value per occurrence is X dollars. If this value is $50 per warning, the present value of peace-of-mind benefits over the 50-year life of a shelter with a 3% real interest rate is $1,325. Figures 5.3 and 5.4 illustrate the impact of peace of mind on the cost per life saved for permanent and mobile homes in the five states above. A peace-of-mind benefit of $50 per household per warning reduces the cost per life saved proportionally (by 44%) in each state, because we assume one warning per year in each state. The cost per life saved in permanent homes is now below $40 million in all five states, with a cost of $20.6 million in Mississippi. Although this repre- sents a much-improved value for shelters, the cost still exceeds the upper range of values of a statistical life observed in markets. For mobile homes,

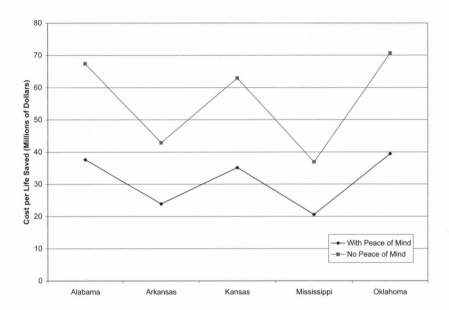

FIGURE 5.3. Peace of mind and the cost-effectiveness of shelters (permanent homes). The peace-of-mind benefit is set at $50 per tornado warning, assuming the household is under a warning once a year.

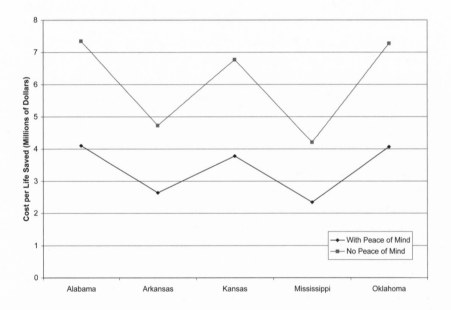

FIGURE 5.4. Peace of mind and the cost-effectiveness of shelters (mobile homes). The peace-of-mind benefit is set at $50 per tornado warning, assuming the household is under a warning once a year.

the cost per life saved is now below $5 million in each state, and the cost in Mississippi is only $2.3 million.

Is $50 a plausible value for a peace-of-mind benefit for the average-sized household? After all, the monetized value of anxiety residents might face during a tornado warning could easily exceed $50, for instance, in a close call where residents can hear the tornado. However, the FAR for warnings is around .75, so in 3 out of 4 warned counties there will be no tornado occurring, and many of the tornadoes that do occur will be some distance from the residents' home and will not constitute a close call. Thus it seems unlikely that the peace-of-mind benefit per warning could exceed $50 per household. The peace-of-mind benefit should plausibly be greater for residents of mobile homes because they face a greater risk of injury or fatality in tornadoes, but residents who choose to live in mobile homes might be less anxious about tornado risk than the average resident.

We can also approach this question from the other end, by specifying the value of a life saved and determining the value of peace of mind per warning required for shelters to break even. For this calculation we will use the inflation adjusted value of a statistical life from EPA (1997) applied for casualties in Chapter 4. So each life expected to be saved by shelters is valued at $7.6 million, and we ask how much peace of mind must be worth during one warning per year to cover the cost. Note that for mobile homes in the five states considered in Figure 5.4, the cost per life saved is less than $7.6 million. Figure 5.5 depicts the break-even peace-of-mind benefit for permanent homes in the five states under consideration in Figure 5.3. The monetized value of fatalities avoided is about $600 in Mississippi and just over $300 in Oklahoma. One warning per year must cover the remaining cost in each case, and with discounting at a 3% real interest rate, the value would need to be about $90 per warning in Mississippi, $94 per warning in Arkansas, and very close to $100 per warning in Alabama, Kansas, and Oklahoma. Again, while these may be perfectly reasonable amounts for close calls, the typical warning is not a close call, and most warnings are after all false alarms. For mobile homes, Tennessee ranked 6th in cost per life saved in Table 5.1, and the break-even value of peace of mind in the Volunteer State is $47 per warning, which as discussed above might be a plausible value. The required value of peace of mind rises to $59 in Indiana and in excess of $70 in Texas. These estimates could be adjusted based on local warning rates, but with the increase in warnings due to Doppler radar and the reductions in warning size due to Storm Based Warnings, it would likely be difficult to estimate the local expected frequency of tornado warnings with precision.

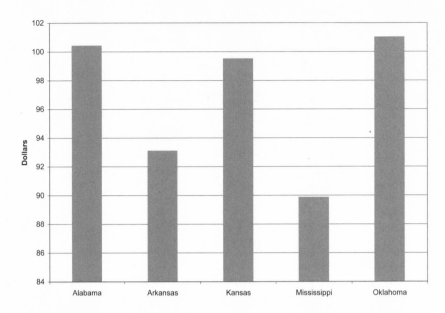

FIGURE 5.5. Break-even values of peace of mind for permanent homes. Per-warning value of peace-of-mind benefit from a shelter in permanent homes with one warning per year, assuming a value of fatalities avoided of $7.6 million.

5.5. A Look Behind the Assumptions of the Tornado Shelter Calculations

Our estimates of the cost-effectiveness of tornado shelters, like any economic calculations, require a number of assumptions. We have detailed the assumptions used in our calculations, like the discount rate, cost, and expected useful life of a shelter. We discuss here some other assumptions, both to highlight them for interested readers and to consider if relaxing these assumptions is likely to overturn our conclusions.

Sharing of shelters. We calculated the cost of shelters on the assumption that a shelter must be installed at each home to prevent all permanent- or mobile-home fatalities. The cost of sheltering therefore is $3,000 per housing unit. However, many underground shelters retailing for around $3,000 can shelter six to ten persons, and the average household size in the United States is less than three persons, so one shelter for every two homes should provide sufficient capacity to shelter all residents. In this case, the cost per household of sheltering and cost per life saved would be cut in half. Some sharing of shelters undoubtedly does occur, and to the extent that shelters can be shared, the figures in Tables 5.1 through 5.4 overestimate the cost per life saved. But not all households are willing to share their shelter. Schmidlin et al. (2009) report that many mobile-home residents surveyed lived within 200 yards of a permanent home or shelter but did not plan to shelter in this location in the event of a tornado. The sharing of shelters by neighbors would introduce some of the complications of community shelters, which we will discuss shortly. Sharing shelters could possibly reduce by half the cost per life saved for permanent homes, but this value still exceeds the upper end of the range of market-revealed values of a statistical life even in the most tornado-prone states. The conclusion that shelters do not appear cost-effective in permanent home is unchanged. In mobile homes, sharing could significantly increase the cost-effectiveness of shelters, as there were 10 states in Table 5.2 with a cost per casualty avoided of between $10 million and $20 million for mobile homes.

Protection for valuables. Our analysis considers only the safety benefits of shelters. However, shelters and particularly safe rooms could also provide protection for valuables and keepsakes, and a safe room might also protect against other natural and man-made hazards, like hurricanes, fire, crime, or terrorist attacks. We do not attempt to quantify these other services shelters might provide (besides the peace-of-mind benefits discussed in Section 5.4), and though including these auxiliary services would improve the value proposition for shelters, it seems unlikely to alter the calculations by more than 10% and would not by itself make shelters cost-effective in permanent homes.

Shelters could alter people's behavior when severe weather threatens. Tornadoes will sometimes strike when residents are not home, and an in-home shelter will not protect residents in these cases. Our calculations based on historical fatalities avoid the not-at-home issue, since the proportion of fatalities in homes (as opposed to other locations) already factors in the likelihood residents are not home when a tornado strikes. The distribution of tornado fatalities by location reflects both the timing of tornadoes (do they strike during the day when residents are likely at work or school) and the likelihood residents are at home when the tornado strikes. What the location of historical fatalities does not control for is the potential for residents who own a shelter to change their behavior when tornadoes threaten. Owning an in-home shelter increases the relative cost of not being home during a tornado, and so people may be more likely to be home during tornadoes after shelter installation. We assume in our calculations that shelters prevent only in-home fatalities and injuries, but by altering household behavior, shelters could also prevent some fatalities in other locations. In this way, potential lives saved could exceed just permanent- or mobile-home fatalities. The activities that might be altered are likely leisure-time activities, and if we look again at Figure 3.14, about three-quarters of fatalities already occur in homes. Some of the remaining fatalities in other locations will not be affected, but suppose that fatalities in vehicles or outdoor locations (which account for 13.6% of fatalities) could be avoided as people alter their behavior. If all these fatalities were avoided, the lives saved by shelters in permanent homes would increase by 13.6%, and the cost per life saved for permanent- home shelters would decline by about 12%, to about $32.5 million in Mississippi. Again, this one factor would not make shelters cost-effective for permanent homes.

Maintenance costs. Our calculations ignore any maintenance required for the shelter to continue to provide complete safety, and any repairs that might be necessary if a tornado strikes the home. Maintenance and repair costs would increase the cost per life saved.

What counts as a "permanent home"? Our permanent-home calculations use the Census Bureau's "1 unit, detached" category, which corresponds most closely with single-family homes. But single-family homes do not include all of the types of housing units that would count as permanent or non-mobile homes, including duplexes, town homes, and apartments. We could make an allowance for these extra units by adding the cost of shelters for these units to the cost for permanent homes, or by deducting from permanent homes the number of fatalities that occur in apartments or other types of units. Surprisingly, very few fatalities appear to occur in these other units, although the fatality location information over the 1996–2007 period does not always state with certainty the type of permanent unit involved. Thus our calculations for permanent homes would probably need little adjustment, and this adjustment would only slightly increase the cost per life saved.

Shelters will not prevent all in-home casualties. We assumed that installation of shelters would prevent all in-home tornado casualties. This assumption simplifies our analysis, but is clearly overly optimistic. Tornado-related casualties will continue for a number of reasons. Some tornadoes occur at night, when residents are asleep and do not receive a warning; residents will sometimes fail to reach their shelter in time; and residents have been injured or even killed while trying to take shelter as a tornado approaches, particularly with underground shelters. In addition, some casualties occur when residents go outside to watch or videotape a tornado instead of taking cover, and shelters clearly cannot prevent these casualties. The data do not allow us to estimate the percentage of in-house casualties that would not be eliminated with any accuracy, but any adjustment for such casualties would increase the cost per fatality or casualty avoided.

Safe rooms versus underground shelters. We have not considered the valuation of safe rooms versus underground shelters. We have used the typical cost of underground shelters ($3,000) in our cost-effectiveness calculations; safe rooms might cost an additional $3,000 or more. As we understand the engineering specifications, underground shelters provide full protection from tornadoes, so the extra cost of a safe room does not directly buy additional safety. However, there may be some indirect benefits of a safe room inside a home as compared with an underground shelter that is located outside. Injuries and even fatalities have occurred as residents have tried to get into an underground shelter. Anecdotal evidence indicates that the number of sheltering-related casualties is not large, though, and seemingly not more than 10% of all tornado casualties, and so the value of underground sheltering casualties avoided will be insignificant compared with the extra cost.[7] In addition, the added convenience of a safe room would need to be quite substantial to justify the extra cost. The convenience of a safe room versus an outdoor underground shelter could be handled in a manner similar to the peace-of-mind benefit. As mentioned in that discussion, residents might expect to be under a tornado warning and thus have to shelter about once per year. At a 3% discount rate, the monetary value of the inconvenience of having to go outside to shelter as opposed to moving to a room in the house would have to be greater than $100 per warning to cover the extra cost of a safe room.

Community shelters. We have not mentioned the possibility of community shelters as an alternative to in-home shelters. Design criteria for community shelters have been issued by FEMA (FEMA 2000), and community shelters are most economical when combined with a multipurpose structure; thus, a building constructed for other purposes would be hardened for use as a tornado shelter. The most attractive application for community shelters would be for mobile-home parks, where many operators currently are providing shelters. Although we do not consider the cost-effectiveness of community shelters directly, we can note several potential problems with them. First is ensuring that the shelters are open for residents whenever tornadoes strike. If shelters are not kept locked, they can become dirty places where residents may not wish to shelter; however, if a shelter is locked and residents caught outside when a tornado strikes, casualties could easily be substantially greater than if residents had sheltered in their homes, especially in permanent homes that offer decent protection. In addition, some residents might wish to shelter even when a tornado warning is not in effect, and if the community shelter is not open these residents could be exposed to thunderstorm hazards (e.g., lightning). Finally, residents would need to leave for the shelter in time to arrive before the tornado strikes. However, if residents can walk or run to the community shelter, "waiting too long" to shelter may not be a problem. Dual-use facilities also have the potential to be in use for other purposes during a warning, and the capacity of the shelter could be exceeded if neighboring residents were to show up. A definitive conclusion would require more detailed study of the issue, but given the high cost of shelters for permanent homes, community shelters would need to reduce the cost per resident by 80% or more to be viable, even if these other complications turned out to be inconsequential.

In sum, we do not think that relaxing the assumptions discussed in this section will affect the bottom line of our analysis: that shelters are cost-effective protection for mobile homes in the most tornado-prone states, but are not cost-effective in permanent homes. Some of the factors discussed in this section would increase our cost per life saved estimates, and others would lower the estimates. We have tried to offer a middle-of-the-road set of assumptions in our analysis. The potential to share shelters and the inclusion of peace-of-mind benefits will definitely improve the value proposition for shelters, particularly for mobile homes in states with tornado risk rankings of 6th to 15th. But even in the most risk-prone states, the cost per life saved in permanent homes would need to be reduced by 80% or more to be within the range of values of a statistical life observed in market studies. Again, this is not meant to imply that the purchase of a shelter by a household living in a permanent home is irrational. Ultimately, the value of tornado safety is subjective, and neither we nor anyone else can say how much this should be worth.

5.6. The Market Value of Mitigation and Tornado Shelters

People have built storm cellars in the Plains states for generations, and in 2000 the National Storm Shelter Association was formed to ensure quality standards in the emerging storm shelter and safe room market. Tornado shelters represent one type of mitigation against natural hazards losses. So how has the market for tornado shelters been performing?

The performance of the market has implications for public policy. In Section 5.2 we described an economic or expected utility framework for durable investment in mitigation like tornado shelters. If residents weigh the expected benefits from mitigation against its cost, if they have reasonably accurate subjective perceptions of the risk (or at least corresponding with our best estimate of the risk), and in the absence of spillover benefits or costs (externalities, to use the economics term), then we can have confidence that the market for protection will work well. We cannot know for sure how many people will want to install tornado shelters because the value of tornado safety is subjective, but households that believe shelters provide sufficient benefits will indeed purchase protection. However, many observers question whether the expected utility framework accurately describes peoples' behavior with respect to natural hazards mitigation, as discussed in Section 5.2. In particular, experts fear that people ignore low-probability, high-consequence events; in the limiting case, people might act as if the probability of a disaster were zero. If people underestimate the risk of a disaster, they will also underestimate the value of mitigation, because mitigation reduces these losses when a disaster occurs. We review here evidence on the performance of the tornado shelter market first to see if residents are ignoring tornado risk altogether, and if not to see if they perceive the risk relatively accurately.

The durable, irreversible nature of the investment in a tornado shelter raises a second issue. Residents likely do not expect to live in a home for the entire useful life of a shelter. We assume a 50-year life for a tornado shelter, while the average American family moves every few years. Imagine a family considers purchasing a shelter with a useful life of 50 years for a home they expect to live in for 5 or 10 years. If a shelter were mobile, it could be packed up and moved along with the family's other possessions. But safe rooms are built into a home and not movable, while an underground shelter would have to be dug up, moved, and buried at the new residence, which is more expensive than just buying a new shelter. Therefore, a tornado shelter would need to be left behind, and although it will protect future residents, the family that makes the initial decision to install a shelter probably does not consider the safety of future residents a high priority. However, the residents who install a shelter can capture these future benefits if they can sell the house at a premium that reflects the value of the shelter to the next residents. The existence of house price differentials for mitigation is thus critically important for a well-functioning market for mitigation. In the absence of price differentials, investment in mitigation will be hampered by a time horizon problem: Any household considering an investment in a shelter will consider only the benefits they expect to realize while living in the home, not the full safety benefits over the useful life of the shelter.

Economists have examined response to various natural hazards risks, and the results have been mixed. These results can provide context for the case of tornadoes. On the one hand, many people fail to insure against floods and earthquakes, suggesting that they ignore these hazards (Kunreuther 1978; Palm 1998; Kunreuther and Pauly 2004). The failure to purchase insurance from the National Flood Insurance Program is particularly telling, as flood insurance is available at less than actuarially fair rates—that is, it is subsidized. Any risk-averse household should want to purchase insurance available at an actuarially fair rate, and this is a rare case where we can make a strong prediction about behavior despite peoples' varying subjective preferences. The demand for subsidized insurance is a direct consequence of risk aversion, and we have abundant evidence that most people are risk averse. Failure to purchase flood insurance is a strong type of evidence for ignoring hazards risk. Evidence also shows that people are unlikely to invest in mitigation, and that home buyers and builders oppose strengthening building codes (Kunreuther 1996). On the other hand, natural hazards risk is often priced in real estate markets, which is consistent with perception of and response to natural hazards. Ceteris paribus, a property that is exposed (or more exposed) to a hazard should be less valuable, because the cost of locating a house or business near a fault line or in a flood plain will be higher. Real estate market price differentials have been observed for earthquakes and floods (Beron et al 1997; Speyrer et al 1991).

Results regarding protection of property and perception of hazards risks must be interpreted with care. The existence of subsidies or the potential for relief after a disaster (from either the government or private charity) can induce people to assume more hazards risk than they otherwise would. For instance, a person who builds a mansion on a barrier island along the Gulf of Mexico might be ignoring the risk of a hurricane, or might be fully aware of the danger but rationally taking advantage of insurance subsidies that lead to taxpayers footing the bill for repairs in the next hurricane. Tornado risk actually provides a good case to test for risk perception and response, because housing choice or installing a storm shelter primarily protects people and not property.

The most relevant evidence for our topic is on the market value of mitigation. In the absence of a premium for mitigation, a time horizon problem will plague the market, and even residents who do not ignore hazards risks might still underinvest in mitigation. Simmons, Kruse, and Smith (2002) found a statistically significant 5% price premium for houses with hurricane blinds in a Texas Gulf coast city. The price premium approximately covered the full cost of hurricane blinds for the average home in their sample.

Several types of evidence suggest that people in general do not ignore tornado risk. Merrell, Simmons, and Sutter (2002) examined applications to the Oklahoma Saferoom Initiative funded by FEMA and the state of Oklahoma after the May 3, 1999, tornadoes, which offered $2,000 rebates on the installation of a tornado shelter or safe room. Specifically, we looked at applications to Phase 3 of the Initiative, which was open to residents from across the state with rebates to be paid based on available funding, whereas Phases 1 and 2 were restricted to homes damaged or destroyed in the May 3 tornado outbreak. A total of 14,000 applications were received, including 11,000 in Phase 3. We found that the incidence of tornadoes in a county increased applications per household from a county, everything else being equal. In addition, Sutter and Stephenson (2008) examined the vote on Oklahoma State Question 696 in the November 2002 general election. The measure offered a credit against property taxes for an in-home safe room or shelter, essentially exempting the area of a shelter from property tax up to 100 square feet, or 10% of the area of a home. The measure passed by a 66%–34% margin statewide and carried a majority in each county of the state. Greater tornado risk increased the share of the vote that the measure received in a county, as well as the number of homes that enrolled for the exemption after passage. A shelter and thus the exemption should be of greater value to residents of more tornado-prone states.

Tornado risk also affects the relative price of mobile homes. Because of the vulnerability that mobile homes face, tornado risk effectively serves as a tax on mobile homes. If residents recognize the risk, then tornado frequency increases the full price of living in a mobile home. Of course tornadoes also create a risk for residents of permanent homes, but because the likelihood of a fatality in a mobile home is 10 times (or greater) the likelihood of a fatality in a permanent home, the impact of tornado risk is relatively greater for mobile homes. Thus, everything else equal, it is relatively more costly to live in a mobile home in a state with a high tornado risk than in a mobile home in a low-risk state. Sutter and Poitras (2010) find that there are fewer mobile homes in high-risk areas, both across states and across counties in Kansas, Oklahoma, and Texas. So again we see that people do not ignore tornado risk. The reduction in the stock of mobile homes is consequential. For instance, a high-risk state like Alabama has an estimated 55,000 fewer mobile homes than it would if the state faced the median level of tornado risk. The reduction in mobile homes due to tornado risk is in line with the reduction that would be predicted based on the value of the risk and estimates of the price elasticity of mobile homes. So it seems that people not only perceive tornado risk, but they appear to perceive it reasonably accurately.

A modest amount of research has also focused directly on tornado shelters. Ozdemir (2005) found that the mean amount people were willing to pay for shelters was $2,450 in a contingent valuation survey of residents of Lubbock, Texas. Ewing and Kruse (2006) found a similar mean willingness to pay: $2,500 in a survey of home buyers in Tulsa, Oklahoma. In addition, Ewing and Kruse found that respondents were willing to pay an extra $600 for a shelter certified by the National Storm Shelter Association, demonstrating the value of this organization and FEMA's efforts to establish performance standards for shelters. Simmons and Sutter (2007a) tested for a market price effect of tornado shelters using data assembled from existing single-family home sales located in Oklahoma County in 2005. About 3% of homes had a tornado shelter, and homes with shelters were available in most price, size, and age ranges, suggesting that buyers could find a home with a shelter as well as other desired characteristics. We found that homes with a shelter sold at a premium of $4,200, or about 3.5% of the median sales price; this is more than enough to cover the price of an in-ground shelter, although less than the cost of a safe room.

Mobile-home parks provide a convenient way to provide community shelters. An economist would predict that mobile-home park operators would provide shelters as an amenity for their residents, similar to swimming pools or laundry facilities. Not all residents will be willing to pay extra for a shelter, so we should not expect all parks to offer shelters, but many presumably would. Schmidlin, Hammer, and Knabe (2001) found in a survey of parks in 11 states that many indeed offered shelters. The percentage of parks in a state with shelters ranged from 12% in Georgia to 76% in Oklahoma and 80% in Kansas. Interestingly, the researchers found relatively low rates of shelters in parks in the Southeastern states, which may be related to the Southeastern component of the mobile-home problem uncovered in Section 3.6. Simmons and Sutter (2007b) conducted a statistical analysis of the rental price of lots in mobile-home parks in Oklahoma using a survey by the Oklahoma Manufactured Housing Association (2005). Fifty-seven percent of parks in the survey reported offering shelters for residents, and 64% of lots in the state were in parks with shelters; in fact, tornado shelters represented the most often provided amenity in Oklahoma mobile-home parks. Furthermore, shelters were widely available across the state: 39 communities in the state had two or more parks, and in 36 of those communities at least one park offered shelters. Lots in parks with shelters rented for about 5% more per month, controlling for other amenities and community characteristics, and this difference in rent would be approximately sufficient to pay for the cost per resident of a multipurpose community shelter based on the cost estimates in FEMA (2000).

5.7. Conclusion

The ability to build shelters and safe rooms capable of allowing residents to survive tornadoes is an impressive engineering feat. We have seen, however, that safe rooms and shelters do not provide cost-effective protection for residents of permanent homes, even in the most tornado-prone states. In other words, the cost per life saved or casualty avoided exceeds the range of values of a statistical life exhibited in market trade-offs. This does not mean that a person purchasing a shelter for their home is making a mistake, because people value safety differently, and shelters can also provide peace-of-mind benefits that enter into residents' valuations. Still, we would not expect widespread market penetration of shelters anytime soon. Shelters may offer a cost-effective way to address the mobile-home problem, however, and we will return to this issue in Chapter 7.

Progress with tornado warnings and warning dissemination actually worsens the value proposition for storm shelters. As tornadoes have become less deadly over time, the number of fatalities shelters can avoid has declined, raising the cost per life saved for shelters. Fatalities per person in the United States have fallen by more than an order of magnitude since the early 20th century, as we saw in Chapter 3. In addition, only about a third of tornado fatalities occur in permanent homes. If not for these changes, fatalities in permanent homes might be 10 times greater than we observe; in this case, the cost per life saved would be about one-tenth (or less) of its current value, which would be close to the range of observed values of a statistical life, at least in the most tornado-prone states. But the NWS's efforts to make tornadoes less deadly also make tornado shelters less cost-effective, or alternatively no longer necessary.

The value proposition of shelters in the future looks even bleaker. Today, some tornadoes go unwarned, or people fail to receive warnings that are issued. The nocturnal and winter tornado vulnerabilities might also be due to difficulties with warning dissemination and response. If progress can be made on these elements of tornado vulnerability, or if mobile homes can be built to survive tornadoes, there will be even fewer deaths for shelters to avoid in the future. On the other hand, the value of safety and thus the value of a statistical life increases with income (Viscusi and Aldy 2003), so rising income in the future will increase residents' willingness to pay to reduce the remaining tornado fatalities. Still, success in improving warnings or reducing casualties weakens the cost-effectiveness of shelters.

5.8. Summary

Engineers have proven that they can design and build structures capable of withstanding even the most powerful tornado winds. Tornado shelters and safe rooms offer the potential to further reduce casualties beyond that already achieved through improved warnings and warning response. The question of whether people should purchase a shelter is ultimately a personal decision depending on how individuals value safety. A separate and trickier question is whether public funds should be used to encourage installation of shelters. Our analysis finds that the cost per life that can be expected to be saved with shelters is rather high for permanent homes, even in the most tornado-prone states. Our analysis does suggest that shelters provide cost-effective protection for mobile homes in the most tornado-prone states.

5.9. Appendix

An individual household contemplating purchase of a tornado shelter would attempt to calculate their expected, discounted utility. This expected utility can be written as follows, on the presumption that all in-home tornado casualties will be eliminated if the household installs a shelter or safe room. A household will want to purchase a shelter if the expected discounted benefits exceed the cost, or

$$\sum_{t=1}^{T} \delta^{(t-1)} * p_{torn} * [p_{fat} * VSL + p_{inj} * VSI] \geq C. \qquad (5A.1)$$

where δ is the household's utility discount factor, p_{torn} is the annual probability of tornado damage at the home in question, p_{fat} and p_{inj} are the expected number of fatalities and injuries in the household, respectively, if a tornado hits the home and they do not have a shelter, VSL and VSI are the values of a statistical life and injury, T is the useful life of the shelter in years, and C is the cost of the shelter, which is assumed to be incurred entirely in year 1.

Our main cost per fatality or casualty avoided calculation uses the VSL or the VSL and VSI for which (5A.1) holds with equality. Instead of applying the inequality to the household, we do so at the state level, and in effect sum the above expression for all permanent- or mobile-home households in the state, and use observed fatalities or casualties for the number of fatalities or casualties in all homes of the type shelters will avoid. In calculating a cost per casualty avoided, we "add" injuries and fatalities at a ratio of 100 injuries to 1 fatality, or substitute VSI = VSL/100 in (5A.1).

We will illustrate the calculations for Mississippi, which has one of the lowest costs for fatality and casualty avoided estimates. Over the period 1950–2007, Mississippi averaged 6.76 fatalities and 97.8 injuries per year. There were 724,757 occupied housing units in the 2000 census category "1, detached" in Mississippi, and 168,520 occupied mobile homes. The numbers reported in columns 1 and 2 of Table 5.1 use the nationwide proportion of fatalities occurring in permanent and mobile homes, .311 and .432, in each state, and with this, Mississippi would experience 2.10 permanent-home and 2.92 mobile-home fatalities per year. With a 3% discount rate, 50-year life of a shelter, and $3,000 cost, the cost per fatality avoided (CPFA) in permanent homes in Mississippi in this case solves

$$(26.502)*(2.10)*CPFA = (724,757)*(3,000), \tag{5A.2}$$

while the cost per fatality avoided for mobile homes is

$$(26.502)*(2.92)*CPFA = (168,520)*(3,000). \tag{5A.3}$$

The adjustment of fatality locations for the state housing stock is made as follows. If the proportion of a given type of home in a state were equal to 0, then the proportion of fatalities in this home would also be 0. We observe the national proportions of permanent homes and mobile homes in the U.S. housing stock, and then construct a weighted average national proportion of permanent and mobile homes using the state housing stock proportions and the percentage of U.S. tornado fatalities over the 1985–2007 period occurring in each state. This yields the following formulas to adjust the proportions of fatalities in permanent and mobile homes in a state:

Proportion(ph) = .482972*(permanent homes/housing units)
Proportion(mh) = 4.096457*(mobile homes/housing units)

If the sum of these adjusted proportions exceeds 1.0, the mobile-home proportion is further adjusted downward to equal one minus the adjusted proportion of permanent-home fatalities. This occurs in four states, and the adjustment in each case is small. The adjusted proportions of fatalities in permanent and mobile homes in Mississippi are .329 and .671.

The estimates in Table 5.3 are generated by substituting permanent- and mobile-home fatalities per year in a state from Table 3.6 into (5A.2) and (5A.3). Mississippi averages .833 permanent-home and .333 mobile-home fatalities per year over 1996–2007.

The estimates based on tornado probabilities in Table 5.4 apply the inequality in (5A.1) directly to a household. The annual probability of tornado damage for a state is taken from Table 2.7, and expected fatalities and injuries per home struck by a tornado, p_{fat} and p_{inj}, from Table 3.11. The cost per fatality avoided in Mississippi calculated in this manner in permanent homes is

$$(26.502)*(.000415)*(.002455) = (3,000),$$

and for mobile homes is

$$(26.502)*(.000415)*(.02166) = (3,000).$$

6

PROPERTY DAMAGE AND COMMUNITY IMPACTS OF TORNADOES

6.1. Understanding the Destructive Impacts of Tornadoes

Research has not examined property damage from tornadoes as extensively as casualties. Several factors are likely in play here. First, the destructive power of tornadoes poses such a threat to life and limb that it is only natural for the main focus to be on casualties. In addition, with winds that can exceed 200 miles per hour and the added force of tornado suction vortices, property damage is seemingly impossible to avoid when a tornado hits. Although a part of a home could be hardened into a safe room capable of surviving an F5 tornado, the construction of homes that are impervious to tornado damage is cost-prohibitive, and building codes do not include wind-load designs for tornadoes. The NWS has instead directed its efforts toward watches and warnings that can potentially save lives. Furthermore, tornadoes simply do not pose the potential for catastrophic losses that affect the functioning of insurance markets and raise homeowners' insurance rates. Tornado events resulting in insured losses in excess of $1 billion (which can include several tornadoes in a large outbreak) are becoming more frequent, but the potential for a $50 billion or $100 billion loss seemingly does not exist. And finally, the available data on property losses suffer from some limitations, which limits the ability of researchers to discern patterns in the losses.

Despite all these limiting factors, this chapter attempts to shed light on how property damage inflicted by tornadoes affects the larger community. We begin by discussing the available data on property damage from tornadoes, and then examine several patterns in damages, including damage by F-scale rating, by year, and across states. Damage provides an alternative way to scale casualties, and yet it does not appear to resolve any of the vulnerabilities identified in Chapter 3. We then undertake a regression analysis of the determinants of damage. Population density and path length increase damage as expected, but the relationship between income and damage appears complicated. And although total damage shows no trend over time, the regression analysis suggests that everything else held constant, tornadoes are causing more damage over time. We conclude by discussing some of the effects of natural hazards on businesses and communities. Tornadoes do not pose the threat to a regional economy that hurricanes or earthquakes do, but are still capable of devastating small towns.

6.2. The Available Data on Property Damage

Our analysis in this chapter uses property losses reported in the Storm Prediction Center (SPC) archive. The method of reporting losses in this archive produces an immediate complication: From 1950 to 1995, damage was reported using a set of eight intervals as described in Table 6.1, while exact damage amounts were reported only beginning in 1996. Therefore, damage was imprecisely reported throughout most of our period of analysis. There are also concerns about the accuracy of reported damage, because of the lack of an incentive for NWS offices to accurately estimate property damage, and due to what appear to be clearly omitted reports for tornadoes that must have caused millions of dollars in damage. Thus our analysis must be interpreted with caution.

Table 6.1 reports the frequency distribution of damage reports for each of the eight intervals used from 1950 to 1995. As can be seen, damage was reported only within intervals of an order of magnitude, indicating a lack of precision. About 29% of tornadoes had zero damage or no report of damage, while only 29 tornadoes (less than .1%) had damage reported in the highest range of $50 to $500 million. Just under 1% of tornadoes had reported damage in the $5 million to $50 million range, and about 5% have reports in the $500,000 to $5 million, $50 to $500, and less than $50 intervals. The interval with the most reports (other than zero or no report) is $5,000 to $50,000, at

TABLE 6.1. NWS Tornado Damage Intervals, 1950–1995

Interval	Damage Range	Percentage of Tornadoes
8	$50 – $500 million	0.08
7	$5 – $50 million	0.95
6	$500,000 – $5 million	4.97
5	$50,000 – $500,000	16.85
4	$5,000 – $50,000	24.77
3	$500 – $5,000	14.44
2	$50 – $500	4.52
1	Less than $50	4.46
0	$0, or No Report	28.97

TABLE 6.2. Distribution of Tornado Damage Reports, 1996–2007

	Nominal Damage	Real Damage
Mean	$769,000	$895,000
Median	$0	$0
Standard Deviation	$11,800,000	$14,000,000
Maximum	$1,000,000,000	$1,244,000,000

Real damage is in 2007 dollars, with inflation adjustment made using the CPI-U index.

nearly 25% of tornadoes. The interval boundaries are not adjusted for inflation, so a tornado producing damage in a given interval in the 1950s actually produced greater inflation-adjusted damage than a similarly reported tornado in 1995.

Table 6.2 describes the distribution of damage reports over the 1996–2007 period. The average damage from a tornado over this period was $769,000 in unadjusted dollars and $895,000 in 2007 dollars adjusted for inflation. The damage distribution is highly skewed, similar to the casualties distribution, as the median tornado over these years had no damage reported, and the standard deviation is over $14 million (adjusted for inflation). The 150 most damaging tornadoes account for 77% of inflation-adjusted property damage. Figure 6.1 presents the distribution of damage reports using the eight intervals from the SPC archive over both the 1950–1995 and 1996–2007 periods to facilitate comparison. The distributions are based on nominal damage (not adjusted for inflation) to maintain consistency with the 1950–1995 reports.

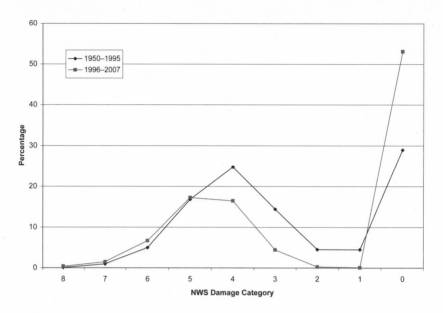

FIGURE 6.1. Distribution of damage reports, 1950–1995 vs. 1996–2007

Tornadoes with no reported damage were more common since 1996 than pre-1996, at 53% of tornadoes versus 29% in the earlier period. This difference is consistent with the improved reporting over time of short-path, weak tornadoes noted in Chapter 2, since these tornadoes are more likely to produce no damage. Fifty-four tornadoes had damage in excess of $50 million in the 12 years since 1996. This is almost double the number of tornadoes reaching this threshold over the 46 years from 1950 to 1995. An additional 1.5% of tornadoes since 1996 had damage in the $5 million–$50 million range; the interval with the largest number of reports since 1996 is $50,000–$500,000. Not surprisingly, damage reports of under $500 have almost disappeared, and proportionally fewer reports in the $500–$5,000 and $5,000–$50,000 ranges occurred since 1996.

We will consider whether a tornado report of no damage is of interest to evaluate the potential for more efficient reporting of weak tornadoes over time, as well as the potential that a zero report is really a missing report. Figure 6.2 reports the number of tornadoes with a positive damage report, and Figure 6.3 reports the percentage of such tornadoes. Figure 6.3 reveals a consistent decline in the percentage of tornadoes with reported damage, from around 85% in the 1950s to around 45% this decade. The percentage declined to around 70% by the mid-1960s and remained steady at this level

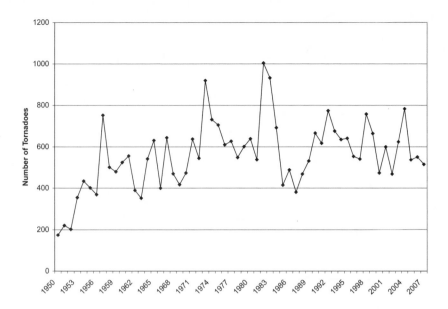

FIGURE 6.2. Tornadoes with positive damage reports, 1950–2007

FIGURE 6.3. Percentage of tornadoes with positive damage reports, 1950–2007

TABLE 6.3. Casualties by Property Damage Category Intervals

Interval	Percentage Killer Tornadoes	Fatalities per Tornado	Injuries per Tornado
8	75.9	12.276	316.48
7	51.3	3.322	54.58
6	17.5	0.607	10.54
5	5.1	0.144	2.12
4	1.6	0.030	0.58
3	0.5	0.007	0.14
2	0.3	0.004	0.16
1	0.3	0.009	0.12
0	0.6	0.035	0.48

Totals are for the years 1950–1995. The property damage category intervals are as described in Table 6.1.

until 1981; then declined to around 60% between 1984 and 1992; and finally declined again to the current level since the early 1990s. Some notable outlier years are also observable, with a percentage of positive reports below the trend in 1954–1956, and only 59% of tornadoes with damage reports in 1963; in the other direction, 83% of tornadoes in 1973 had positive damage, as did over 95% of tornadoes in 1982 and 1983. The number of tornadoes with positive damage has averaged about 600 per year since 1975, with no apparent trend. This pattern is consistent with the substantial underreporting of tornadoes in the 1950s, with the reported tornadoes in the 1950s and 1960s generally being quite significant; compare this with many more tornadoes producing no property damage in recent decades.

Presumably, tornadoes that result in fatalities and injuries will also cause property damage. Casualties result from an interaction of tornadoes with people, which generally means tornadoes impacting the built environment. Although casualties can occur in the outdoors, away from buildings, the overwhelming proportion of casualties occur when tornadoes strike populated areas, thus also producing property damage. Table 6.3 reports the percentage of killer tornadoes as well as fatalities and injuries per tornado over the 1950–1995 period by damage category intervals. As we would expect, casualties occur much more frequently in the higher damage categories. Tornadoes with damage in excess of $50 million had an average of 12 fatalities and 316 injuries; these averages exceed casualties per tornado in the under $50 damage interval by more than three orders of magnitude. Casualties drop steadily until category 3 (damage less than $5,000), where the numbers begin to flatten out. Just over half of tornadoes in category 7 produced a fatality, and these tornadoes resulted in 3 fatalities and 55 injuries on average. The percentage of killer tornadoes falls to 18% (category 6), then 5% (category 5), and 1.5% (category 4), with similar declines in casualties per tornado. Only about half of one percent of tornadoes in the 3, 2, 1, and 0 categories produced fatalities, which is less than 1 fatality for every 100 tornadoes in these categories. The most surprising pattern from Table 6.3 is the substantial number of tornadoes that result in casualties but no reported damage. A total of 61 killer tornadoes that resulted in over 350 fatalities and nearly 5,000 injuries have a 0 category report in the SPC archive. The tornadoes with no damage report include a February 1971 F4 tornado in Mississippi that killed 58 persons and injured 795 (the last tornado in the United States to kill 50 or more persons), as well as eight other F4 or F5 tornadoes that killed 10 or more persons. Thus, in addition to the loss of precision due to the interval estimates, there are clearly tornadoes with missing damage estimates. We see some evidence that the killer tornadoes in category 0 are likely missing a damage report, as fatalities and injuries per tornado in this category are about four times greater than in categories 1, 2, and 3. All of our analysis of damage must be viewed with serious caution. In fact, a warning label probably should be affixed to virtually any natural hazards property damage analysis using publicly available data sets, as noted by Gall, Borden, and Cutter (2009).

We wish to conduct some analysis using damage amounts, so we need to convert the damage intervals into damage amounts. We do this by assigning to each tornado the dollar value of the midpoint of the interval. Alternatively, it would be possible to use the upper or lower bound of the interval; the University of South Carolina data set uses the lower value of each interval in its damage estimates (Gall, Borden, and Cutter 2009), but this is unnecessarily conservative. Another alternative would be to attempt to estimate the underlying statistical distribution consistent with the proportion of observations in each interval and calculate an expected value in each range based on the underlying density function. However, given the imprecision of the intervals and the other limitations of the damage reports, the mean of the interval end points will suffice for our purposes. We then adjust for inflation using the national Consumer Price Index for all urban consumers, so our real damage totals are in 2007 dollars. For tornadoes since 1996, we use the reported damage figures and adjust for inflation. Note that we do not scale for population, wealth, or income in our dollar amounts, as do Brooks and Doswell (2001), and as is often done in the literature of societal impacts of weather (Pielke and Landsea 1998; Pielke et al. 2008).

6.3. Patterns in Tornado Damage

We now consider the costliest tornadoes to occur since 1950. The SPC archive is of little value in ranking tornadoes occurring prior to 1996, as 29 tornadoes fall into the category 8 designation. Instead, we use the damage amounts reported by Grazulis (1993, 1997) for these tornadoes, and then the damage estimates from the SPC archive for tornadoes since 1996. Table 6.4 reports the tornadoes with real damage (in 2007 dollars) in excess of $400 million since 1950. Our list is based on an inflation adjustment only, not any other normalization of damage. The table also reports damage (in millions) per mile of path length of the tornado, in addition to the F-scale rating of the tornado. Damage per mile gives an idea of just how destructive tornadoes can be. The 1999 Oklahoma City F5 tornado ranks 1st in total damage at over $1.2 billion, while the 1979 Wichita Falls, Texas, F4 tornado ranks 2nd at $1.1 billion. Coincidentally, both tornadoes were equally destructive per mile of path length, producing about $33.5 million of damage per mile. Five other tornadoes exceed $500 million in inflation-adjusted damage. The 1970 Lubbock, Texas, F5 tornado ranks 3rd at $722 million, and was the most destructive storm for its path length at nearly $86 million per mile—70% greater than for the next most destructive tornado in Table 6.4. One tornado from the 1974 super tornado outbreak makes the list—the Xenia, Ohio, F5 tornado, which ranks 7th—while the 1953 Worcester, Massachusetts, F4 tornado ranks 9th at $405 million. Due to the imprecise and incomplete nature of the damage estimates, especially prior to 1996,[1] this list should be considered merely suggestive. A more extensive ranking of U.S. tornadoes by damage is available from Brooks and Doswell (2001).

We next compare damage by F-scale rating. As we would expect, damage increases with the F-scale rating of the tornado, as Table 6.5 indicates. About 90% of tornadoes rated F2 or stronger have a report of damage greater than zero, although interestingly the percentage is approximately equal for the F2 through F5 categories. Damage reports fall to 81% for F1 tornadoes, and only 38% of F0 tornadoes have a positive damage report. Half of tornadoes with missing F-scale ratings have positive damage reports, so again we see that these tornadoes appear to resemble F0 tornadoes with a few F1 tornadoes mixed in. Undoubtedly, many of the strong and violent tornadoes with no damage reports did in fact produce property damage that simply failed to be reported and recorded in the SPC archive. Although the percentage of tornadoes with positive damage reports does not increase above the

TABLE 6.4. Costliest Tornadoes, 1996–2007

State	Year	Cost	Cost per Mile	F-Scale
Oklahoma City, Oklahoma	1999	$1,244	$33.6	5
Wichita Falls, Texas	1979	$1,142	$33.5	4
Lubbock, Texas	1970	$722	$85.9	5
Topeka, Kansas	1966	$640	$30.3	5
Hartford, Connecticut	1979	$571	$50.5	4
Athens, Georgia	1973	$528	$7.3	2
Xenia, Ohio	1974	$420	$13.4	5
Moore, Oklahoma	2003	$417	$24.1	4
Worcester, Massachusetts	1953	$405	$11.6	4

Cost is in millions of 2007 dollars, with inflation adjustment using the CPI. Cost estimates for tornadoes prior to 1996 with damage in the $50–$500 million category in the SPC archive are taken from Grazulis (1995, 1997), when available.

TABLE 6.5. Damage by F-Scale Rating

F-Scale	Percentage with Damage > 0	Damage per Tornado
5	89.06	174.30
4	92.41	49.80
3	92.04	10.49
2	89.28	2.32
1	81.19	0.54
0	38.34	0.04
−9	50.30	0.21

Damage is in millions of 2007 dollars.

F2 category, damage does escalate significantly with each F-scale category. Damage averaged $174 million for F5 tornadoes, more than three times the average of $50 million for F4 tornadoes, while damage averaged $10 million for F3 tornadoes and $2.3 million for F2 tornadoes. F1 tornadoes resulted in an order of magnitude greater damage than F0 tornadoes, although almost half of the difference is due to the greater frequency of reported damage; for tornadoes with damage reports, damage was six times greater for F1 than F0. Overall, the strength of a tornado, as proxied by the F-scale rating, greatly affects damage, as the average reported damage for an F5 tornado is over 4,000 times greater than damage for an F0 tornado. Recall that fatalities

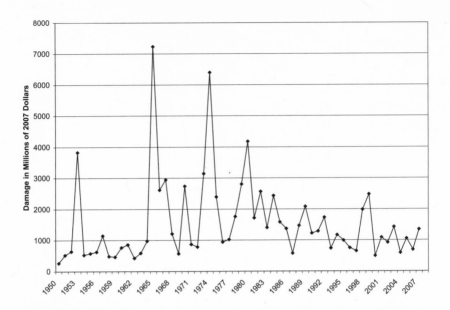

FIGURE 6.4. Tornado damage by year. Damage amounts in 2007 dollars, adjusted using the CPI. Damage amounts for tornadoes prior to 1996 are inferred based on the midpoint of damage intervals.

per tornado and injuries per tornado were 15,000 and 5,000 times greater, respectively, for F5 tornadoes than F0 tornadoes (see Table 3.5), so damage actually increases slightly less with F-scale rating than casualties.

We turn next to the distribution of damage over time. Figure 6.4 displays total real damage by year since 1950. Over the period, damage averaged $1.55 billion per year and median damage was $1.11 billion, with substantial year-to-year variation. The year with the greatest damage was 1965 at an estimated $7.2 billion, followed by 1974 at $6.4 billion, 1980 at $4.2 billion, and 1953 at $3.8 billion. Given the imprecision of individual storm estimates and the virtual certainty of missing damage reports, we cannot state with much assurance that damage in 1965 actually exceeded damage in 1974; nonetheless, these four years do stand out as the worst years for tornado damage. Annual damage was less than $1 billion in 26 of the years, or almost half of all the years in the data set, with a minimum of $263 million in 1950. Over the entire period, total damage exhibits essentially no time trend; a simple linear regression shows a slight decrease of $3 million per year, but the estimate does not approach statistical significance. Close inspection of Figure 6.4 indicates three periods of damage in the annual totals. Over the first period, from 1950 to 1964, reported damage averaged $845 million per year and was less than $1 billion in 13 of 15 years. The average over these years is boosted by the damage total in 1953. Between 1965 and 1989, reported damage averaged $2.3 billion per year; the three years with the largest damage totals occurred in this period, and damage was less than $1 billion only five times. The third period, from 1990 to 2007, includes the years with damage estimates as opposed to intervals. In this period, damage averaged $1.1 billion per year and exceeded $2 billion in a year just one time, in 1999; the following year, damage was only $500 million. The variance of year-to-year damage is smaller in this period as well. As mentioned, these loss figures are imprecise, but we see no evidence of a time trend in damage.

Rising natural hazards losses have been a concern of policy makers and insurance companies in recent years. We see that overall tornado damage has not been increasing, and in contrast to hurricanes, none of the five worst damage years has occurred in the last 20 years. Inferences regarding trends in losses over time depend on the level of damage attributed to each interval for tornadoes prior to 1996; we have applied the midpoint, but as mentioned previously, other researchers have applied the lower bound of the interval, which underestimates the damage in the years prior to 1996. If we generate a time series of damage using these lower bounds, then 1965 ranks as the 5th highest annual damage total (behind 1999, 1998, 2003, and 2007), and we obtain a statistically significant upward trend in annual damage of $15 million per year. If we instead use the lower quarter point of the interval (the average of the lower bound and midpoint) to generate the damage series, 1965 and 1974 return to the 1st and 2nd rankings, 1999 falls to 3rd, and we obtain a statistically insignificant $5 million per year upward trend in damage.

The annual damage totals illustrate how tornadoes provide little danger of catastrophic losses that might potentially disrupt insurance markets. Six hurricanes have resulted in insured losses in excess of the worst annual damage total for tornadoes in Figure 6.4, and all six of these worst-loss hurricanes have occurred since 1992. Losses well in excess of those observed in most tornadoes are certainly possible with a long-track urban tornado (Wurman et al. 2008) and as illustrated with normalized damage from the 1896 St. Louis tornado (Brooks and Doswell 2001); but in general, tornadoes are close to a normal type of hazard for the insurance industry.

We turn next to damage across states. Table 6.6 reports total damage by state over the period 1950–2007, in millions of 2007 dollars. The table also reports the percentage of state tornadoes with reports of positive of damage to allow us to consider potential patterns across the country in the reporting of damage. Total state damage is scaled in two ways, per state tornado (in millions of dollars) and per capita (in dollars) based on the average state population in the 1950 through 2000 censuses.

The percentage of state tornadoes with no damage report varies widely, with no apparent simple pattern. Vermont, a low-tornado-rate state, has the highest percentage of tornadoes with reported damage greater than zero, at 97%, and several other states with low tornado rates also have high percentages of positive damage reports, including Massachusetts, New York, and New Hampshire. At the same time, some higher tornado rate states also have high percentages of positive damage reports, like Georgia, Alabama, and Tennessee, while other Northeastern states like Connecticut and Rhode Island have low percentages of positive damage reports. Nevada has the lowest tornado rate, and also the lowest percentage of tornadoes with positive damage. Like Nevada, the other states with the lowest percentages of positive damage reports also have low population densities, like Colorado, New Mexico, North Dakota, and South Dakota (particularly the eastern parts of Colorado and New Mexico, which are most often struck by tornadoes). Thus we see some evidence that tornadoes striking rural areas are less likely to result in damage, even if observed and reported (see Chapter 2).

The states with the most tornadoes over the period also tend to have the most damage. Texas ranks first in damage at $9.3 billion, followed by Indiana ($7.0 billion), Oklahoma ($6.3 billion), Georgia ($5.3 billion), and Kansas ($5.2 billion). Twenty-four states had damage in excess of $1 billion in these 58 years. At the other end of the distribution, Nevada had only $6 million in damage over the period, followed by Rhode Island ($7 million), Vermont ($13 million), Delaware ($23 million), and Idaho ($23 million). Scaled damage tells a different story, as we might expect. Damage per tornado is greatest in Connecticut, at $17 million per tornado, followed closely by Massachusetts at $16 million per tornado. To place the averages for these two states in context, damage per tornado for all tornadoes in our data set was $1.8 million. Connecticut and Massachusetts face only a moderate number of tornadoes, but tornadoes in these states tend to strike relatively populated areas and produce considerable damage. Indiana ranks 3rd at $6 million per tornado, followed by Ohio ($4.8 million) and Georgia ($4.2 million). Nevada ranks last at $80,000 per tornado, and other mountain states have the next lowest levels of damage per tornado, due to a combination of mostly weak tornadoes and sparse populations.

Damage per capita, as reported in the last column of Table 6.6, indicates the impact of property damage on the residents and economy of a state. For a given amount of damage, a larger state population represents a larger base

TABLE 6.6. Tornado Damage by State

State	Percentage with Damage > 0	Total Damage	Damage per Tornado	Damage per Capita
Alabama	89.2 [5]	$2,771 [14]	$1.866 [16]	$750 [11]
Arizona	49.0 [39]	$214 [34]	$1.019 [32]	$84 [35]
Arkansas	62.2 [30]	$2,286 [19]	$1.602 [22]	$1,061 [8]
California	62.4 [29]	$244 [32]	$0.674 [36]	$11 [46]
Colorado	27.0 [47]	$288 [31]	$0.166 [46]	$110 [33]
Connecticut	75.3 [23]	$1,395 [24]	$17.221 [1]	$482 [16]
Delaware	78.9 [18]	$22.5 [45]	$0.395 [39]	$40 [39]
Florida	77.3 [2]	$2,823 [13]	$0.960 [33]	$319 [25]
Georgia	93.7 [2]	$5,265 [4]	$4.206 [5]	$984 [10]
Idaho	55.7 [33]	$23 [44]	$0.126 [47]	$26 [45]
Illinois	55.3 [35]	$3,431 [10]	$1.758 [18]	$316 [26]
Indiana	79.8 [17]	$7,019 [2]	$6.015 [3]	$1,363 [4]
Iowa	75.0 [25]	$3,438 [9]	$1.676 [20]	$1,226 [6]
Kansas	47.2 [41]	$5,206 [5]	$1.585 [23]	$2,253 [1]
Kentucky	84.4 [11]	$1,610 [23]	$2.457 [9]	$469 [17]
Louisiana	81.0 [15]	$2,357 [18]	$1.563 [25]	$629 [15]
Maine	80.0 [16]	$31.3 [43]	$0.313 [42]	$29 [43]
Maryland	86.9 [10]	$447 [29]	$1.631 [21]	$113 [32]
Massachusetts	91.4 [4]	$2,399 [17]	$15.890 [2]	$428 [18]
Michigan	82.4 [14]	$3,427 [11]	$3.770 [6]	$399 [20]
Minnesota	47.4 [40]	$4,386 [6]	$3.133 [8]	$1,116 [7]
Mississippi	83.6 [12]	$2,967 [12]	$1.859 [17]	$1,227 [5]
Missouri	76.4 [21]	$3,464 [8]	$1.990 [12]	$727 [12]
Montana	41.6 [42]	$101 [41]	$0.279 [43]	$136 [30]
Nebraska	57.8 [32]	$2,637 [15]	$1.095 [31]	$1,742 [3]
Nevada	22.7 [48]	$6.0 [48]	$0.080 [48]	$7 [48]
New Hampshire	87.7 [7]	$33.9 [42]	$0.418 [38]	$39 [40]
New Jersey	71.0 [27]	$211 [35]	$1.532 [27]	$31 [42]
New Mexico	39.0 [45]	$118 [39]	$0.232 [44]	$97 [34]
New York	88.0 [6]	$827 [26]	$2.356 [10]	$48 [38]
North Carolina	75.0 [24]	$1,687 [21]	$1.687 [19]	$295 [27]
North Dakota	37.7 [46]	$422 [30]	$0.342 [41]	$666 [15]
Ohio	92.0 [3]	$4,101 [7]	$4.775 [4]	$401 [19]
Oklahoma	65.7 [28]	$6,264 [3]	$2.033 [11]	$2,245 [2]
Oregon	54.7 [36]	$184 [36]	$1.941 [14]	$78 [36]
Pennsylvania	78.0 [19]	$2,462 [16]	$3.742 [7]	$212 [28]
Rhode Island	55.6 [34]	$7.2 [47]	$0.793 [35]	$8 [47]
South Carolina	71.6 [26]	$956 [25]	$1.201 [29]	$324 [24]
South Dakota	39.0 [44]	$681 [27]	$0.436 [37]	$986 [9]
Tennessee	87.0 [9]	$1,613 [22]	$1.923 [15]	$373 [22]
Texas	50.1 [38]	$9,290 [1]	$1.232 [28]	$692 [13]
Utah	52.7 [37]	$219 [33]	$1.957 [13]	$163 [29]
Vermont	97.2 [1]	$13 [46]	$0.360 [40]	$27 [44]
Virginia	87.5 [8]	$612 [28]	$1.145 [30]	$120 [31]
Washington	59.6 [31]	$146 [37]	$1.550 [26]	$37 [41]
West Virginia	83.6 [13]	$104 [40]	$0.893 [34]	$56 [37]
Wisconsin	75.7 [22]	$1,772 [20]	$1.580 [24]	$397 [21]
Wyoming	39.7 [43]	$129 [38]	$0.229 [45]	$326 [23]

Ranks in parentheses. Total damage and damage per tornado in millions of 2007 dollars.

to spread the damage across, thus reducing the impact. The top five states in damage per capita are Kansas ($2,253 per person), Oklahoma ($2,245), Nebraska ($1,742), Indiana ($1,363), and Mississippi ($1,227); the Plains states at the heart of Tornado Alley have the greatest damage per person, reflecting these states' high rates of tornado activity and relatively modest populations. However, even in Kansas and Oklahoma, the overall level of damage per capita is still modest, at about $39 per person per year, or about .1% of per capita income annually. Thirty-three states experience tornado damage of less than $10 per person per year, and Nevada again ranks last at $7 per person, so damage per capita varies across states by a factor of just over 300. We see the effect of a large population in diffusing the impact of tornadoes in Texas, Ohio, and Illinois: these states rank 1st, 7th, and 10th in total damage, but due to their large populations only 13th, 19th, and 26th in damage per capita. Illinois's total of $3.4 billion in damage appears quite burdensome, but works out to just over $5 per capita per year, or about the cost of lunch at a fast food restaurant.

6.4. Damage and Casualties

Our casualty analysis used tornado-path demographic variables that were constructed based on counties struck. County population density was intended to control for whether a tornado struck a populated area or not. As discussed in Chapter 3, counties are large relative to tornado damage paths, and in many counties population is clustered in a few towns or cities, so persons per square mile for a county as a whole can be a misleading measure for the population of the typical square mile of land area. Despite this limitation, however, county population density performed well in the regression analysis. The damage figures provide an alternative means of examining casualties. Damage, in theory, should control for the number of structures damaged or destroyed in a twister. A tornado of a given F-scale and length that struck a county with low population density but did considerable damage likely devastated one or more towns, and the potential for casualties would be greater than indicated by the density.

Damage and casualties are correlated, as we would expect. Table 6.7 reports the percentage of tornadoes with positive damage reports and average real damage per tornado for killer and injury tornadoes, compared in each case with tornadoes that did not kill or injure anyone. Overall, 93% of killer tornadoes had a positive damage report, compared with 63% of non-killer

TABLE 6.7. Tornado Casualties and Damage

	Percentage with Damage > 0	Average Damage
Killer Tornadoes	93.4	$37 million
Non-Killer Tornadoes	63.1	$860,000
Injury Tornadoes	93.8	$11 million
Non-Injury Tornadoes	59.3	$380,000

Damage is adjusted for inflation in 2007 dollars.

tornadoes. Killer tornadoes produced on average $37 million of reported property damage, compared with $860,000 for non-killer tornadoes. In the same vein, 94% of injury-producing tornadoes had a positive damage report compared with 59% of non-injury-producing tornadoes, and damage averaged $11 million in injury tornadoes and $380,000 in non-injury-producing tornadoes, so we see that tornadoes that did not injure anyone had on average a modest impact on property. The proportions of killer and injury tornadoes with positive damage reports were about equal and in fact slightly lower for the 1996–2007 years, indicating that a lack of damage reports for casualty-generating tornadoes (which likely indicates missing data and not $0 in damage) is not confined to the early part of the period. Overall the correlation between fatalities and damage is +.38 and between injuries and damage is +.59.

Casualties scaled by damage provide another way to investigate differences in tornado lethality across states. The most tornado-prone states will also have the most damage, but if damage provides a better measure of the impact of tornadoes on society, casualties per million dollars of damage may not vary across states. Table 6.8 presents fatalities and casualties (fatalities + injuries) per million dollars of damage over the 1950–2007 period, for the top 15 states by each measure. Fatalities per million dollars of damage reveals the same Southeastern component as our fatalities analysis in Chapter 3. The top four states are Tennessee, at one fatality for every $6 million in damage, followed by Arkansas, Mississippi, and Alabama, with one fatality for every $7.7 million. Kentucky and Louisiana also rank in the top 10, while North Carolina, Florida, and South Carolina are in the top 15. As for the Tornado Alley states, Texas ranks 11th, and Kansas, Oklahoma, and Nebraska do not make the top 15, suggesting that tornadoes at the heart of Tornado Alley are not especially deadly for their impact on the built environment. Several low-fatality states feature prominently among the top states in casualties per million in damage, including Delaware and Rhode Island in the 1st and 2nd

TABLE 6.8. Tornado Casualties per Million Dollars of Damage

Fatalities per Million Dollars of Damage		Casualties per Million Dollars of Damage	
1 Tennessee	0.167	1 Delaware	3.335
2 Arkansas	0.147	2 Rhode Island	3.222
3 Mississippi	0.132	3 Tennessee	2.429
4 Alabama	0.130	4 Arkansas	2.204
5 Delaware	0.089	5 Washington	2.121
6 Kentucky	0.072	6 Alabama	2.091
7 Michigan	0.071	7 Mississippi	2.044
8 Louisiana	0.063	8 Kentucky	1.768
9 Illinois	0.059	9 South Carolina	1.382
10 Missouri	0.059	10 New Mexico	1.355
11 Texas	0.058	11 North Carolina	1.339
12 North Carolina	0.058	12 Illinois	1.239
13 North Dakota	0.057	13 Florida	1.217
14 Florida	0.057	14 Louisiana	1.158
15 South Carolina	0.056	15 Ohio	1.116

slots, with Washington and New Mexico at 5th and 10th, respectively. Rhode Island did not have a tornado fatality over the period, but had more than 3 injuries per million dollars of damage. However, with only $7 million in damage, the lack of fatalities in Rhode Island may have been due to the relatively small total amount of damage over the period, and not a particularly low lethality of tornadoes. It is conceivable that Rhode Island could in fact have the same rate of fatalities per million dollars of damage as Tennessee, and simply by luck not have experienced a fatality over the period; indeed, there is about a 1 in 4 chance that Tennessee's rate of .167 fatalities per million dollars of damage would not result in a fatality given Rhode Island's $7 million in observed damage. The Southeastern component of tornado vulnerability is apparent in casualties per million dollars of damage as well, as Tennessee, Arkansas, Alabama, Mississippi, Kentucky, South Carolina, Florida, and Louisiana all rank in the top 15 for scaled casualties.

Damage does not appear to resolve the differences in lethality observed across the nation. The correlations between fatalities and casualties per million dollars of damage and the state casualty index derived in Chapter 3 for the states with an index value (see Table 3.10) are positive but modest, at +.12 and +.13, respectively. Damage may allow a better control for the number of buildings actually in a tornado's path, but Southeastern states appear to be more vulnerable even when controlling for human interaction with the built environment.

6.5. The Determinants of Tornado Damage

So far in this chapter we have examined patterns of tornado damage. We now turn to an examination of the determinants of tornado damage, similar to our analysis of the determinants of casualties. Which factors lead to greater damage from a tornado? We approach the question using regression analysis and the same control variables from our casualty analysis. Once again, the reader is cautioned that the analysis can be no more informative than the quality and accuracy of the damage reports.

The imprecision of the pre-1996 damage reports due to the use of damage intervals, the potential for missing reports, and the lack of an incentive for the NWS to invest to improve the accuracy of damage estimates put into question whether it's worth analyzing the reported damage of a tornado versus its actual damage (in 2007 dollars). As discussed earlier, the damage amounts we apply for the 1950–1995 period use the midpoint of the intervals in Table 6.1. Our econometric analysis is complicated by the fact that tornadoes with nonzero damage reports are a non-random sample of all tornadoes. To address this problem, we employ the two-stage sample selection model pioneered by Nobel Prize–winning econometrician James Heckman (1976; Greene 2000). The appendix to this chapter discusses the sample selection model in greater detail, but the first stage is a probit regression of whether a tornado had a nonzero damage report, and the second stage is an ordinary least-squares analysis of the natural log of reported damage for tornadoes with positive damage.

We estimate four sets of models. We begin with a model for the entire 1950–2007 period, and then estimate separate specifications using 1950–1995 and 1996–2007 tornadoes to control for any difference in patterns resulting from the change from damage intervals to dollar amounts. Finally, we estimate models using 1996–2007 tornadoes and additional demographic and economic control variables. We exclude the time of day, weekend, and month variables from our models, since although timing can plausibly affect human alertness and vulnerability, there seems little reason to expect a tornado of given strength to do more damage after dark or in the winter months, for example.

TABLE 6.9. First-Stage Probit Regressions of a Positive Damage Report

	1950–2007	1950–1995	1996–2007	1996–2007
Intercept	2.98 (.167)**	2.79 (.195)**	−7.58 (.442)**	−13.94 (.962)**
Pop Density	.181 (.005)**	.199 (.006)**	.192 (.009)**	.142 (.014)**
Income	−.377 (.030)**	−.381 (.036)**	−.254 (.064)**	.569 (.138)**
Time Trend	−.017 (.001)**	−.001 (.001)	.170 (.004)**	.170 (.004)**
F1	.851 (.016)**	.900 (.019)**	.768 (.030)**	.754 (.030)**
F2	1.08 (.023)**	1.12 (.027)**	1.03 (.048)**	1.02 (.049)**
F3	1.20 (.042)**	1.12 (.050)**	1.50 (.088)**	1.49 (.088)**
F4	1.12 (.081)**	.974 (.091)**	1.54 (.193)**	1.57 (.193)**
F5	.823 (.216)**	.582 (.217)**	5.32 (34.09)	5.42 (33.42)
Path Length	.090 (.004)**	.119 (.006)**	.139 (.009)**	.135 (.009)**
Mobile Homes				.007 (.025)
Rural				.063 (.020)**
Non-white				.078 (.017)**
Male				−1.65 (.419)**
Under 18				−.563 (.149)**
Over 65				−.066 (.070)
Commute				.306 (.036)**
No High School				.022 (.059)
College				−.112 (.049)**
Home Age				.091 (.059)
Poverty				.248 (.066)**
Log Likelihood	−24751	−15645	−7069	−6928

The dependent variable is a dummy variable that equals 1 for tornadoes with a positive damage report. The coefficients are the probit coefficients with standard errors in parentheses. ** and * indicate significance at the .05 and .10 levels, respectively.

Table 6.9 presents the first-stage probit models of whether a tornado produced positive damage. All of our control variables are significant in the 1950–2007 specification. As we would expect, a more densely populated storm path significantly increases the probability of damage, since a tornado is more likely to strike buildings in a more densely populated county. Path length also increases the probability of positive damage, also as expected. The time trend is negative and significant, consistent with the percentage of tornadoes with positive damage by year depicted in Figure 6.1. In more recent years, more short and weak tornadoes are being reported, and these tornadoes are less likely to produce damage, or at least a damage report. The

F-scale dummy variables are all significant and increase the probability of a damage report relative to an F0 tornado, as we saw in the breakdown of damage by F-scale in Table 6.5. Interestingly, the point estimate for F5 tornadoes is actually smaller than for F2, F3, or F4 tornadoes and approximately equal to an F1 tornado. We saw in Table 6.5 that about 90% of tornadoes rated F2 or stronger had positive damage reports, and F5 tornadoes on average have longer damage paths, which increases the probability of damage directly. The most surprising result in the probit regressions concerns income: Higher income reduces the probability of a damage report. We would expect that damage and income should be positively related, as wealthier communities have more possessions and higher-valued property. Of course, here we are just considering the probability of damage greater than zero, which might weaken our intuition; despite that, wealthier communities should invest more in emergency management, have better insurance, and be at least as likely to attract NWS damage survey teams, so the negative sign on income is still unexpected.

Columns 2 and 3 of Table 6.9 present the probit models estimated for tornadoes that occurred through 1995 and since 1996, respectively, to see if the change in the method of reporting damage affects our inferences. Most of the results in the two time periods are robust, meaning the variables have similar estimate impacts in each sample. The one notable difference is the time trend, which is not unexpected. No time trend is observed over the 1950–1995 period; although the percentage of tornadoes with damage reports declined over these years, as Figure 6.3 illustrates, when we control for F-scale (more F0 tornadoes occur) and path length, we see no additional effect of time on the probability of damage. However, since 1996, we observe a statistically significant increase in damage reports over time. We also see a large point estimate for F5 tornadoes since 1996, which reflects the fact that all tornadoes rated F5 since 1996 had positive damage.

Column 4 of Table 6.9 reports a specification using 1996–2007 tornadoes and includes other demographic control variables. We have no strong expectation regarding most of these variables, and they are included to test whether the determinants of damage reports are robust to inclusion of these additional variables. The population density, time trend, path length, and F-scale relationships are robust, as the coefficients for these variables have the same signs, significance, and marginal effects as in column 3. Income changes sign, however, to positive and significant, which conforms to expectations. Note that several of the added control variables correlate with income (specifically, college is negative and significant), and this may be

TABLE 6.10. Second-Stage OLS*** Regression on Inflation-Adjusted Damage

	1950–2007	1950–1995	1996–2007	1996–2007
Intercept	−6.08 (.295)**	−5.11 (.335)**	−13.38 (2.04)**	−18.59 (3.46)**
Pop Density	.341 (.020)**	.372 (.026)**	.466 (.043)**	.339 (.038)**
Income	−.439 (.059)**	−.659 (.074)**	−.280 (.127)**	.460 (.271)*
Time Trend	.111 (.002)**	.121 (.002)**	.187 (.035)**	.163 (.033)**
F1	1.95 (.110)**	2.09 (.144)**	2.72 (.177)**	2.56 (.162)**
F2	3.26 (.126)**	3.35 (.164)**	4.62 (.227)**	4.42 (.210)**
F3	4.37 (.139)**	4.44 (.167)**	6.21 (.301)**	5.97 (.281)**
F4	5.62 (.157)**	5.72 (.182)**	7.21 (.372)**	6.98 (.360)**
F5	6.86 (.307)**	6.84 (.337)**	10.37 (.804)**	10.12 (.786)**
Path Length	.371 (.021)**	.376 (.015)**	.461 (.033)**	.447 (.031)**
Mobile Homes				.188 (.045)**
Rural				−.199 (.039)**
Non-white				−.045 (.035)
Male				−1.82 (.852)**
Under 18				−.434 (.315)
Over 65				−.320 (.135)**
Commute				.286 (.092)**
No High School				.050 (.109)
College				−.096 (.089)
Home Age				.215 (.115)*
Poverty				.268 (.129)**
IMR	−.062 (.207)	−.040 (.292)	2.20 (.342)**	1.95 (.321)**
Adjusted R2	.5194	.4749	.5306	.5365

The dependent variable is the log of damage, and only tornadoes with positive damage are included in the regression. All of the control variables are entered in natural logs except the time trend and F-scale dummy variables. ** and * indicate significance at the .05 and .10 levels, respectively. *** "OLS" is Ordinary Least Squares, a basic regression technique.

capturing the negative effect previously attributed to income. A larger rural population increases the likelihood of positive damage; in a county with a given overall population density, a larger rural population means that people and buildings are more likely to be in any tornado's path, leading to at least some damage. Since rural counties have lower incomes, everything else equal, rural may also be capturing some of the effect previously attributed to income. The other significant determinants of positive damage include a larger nonwhite population, a larger female population, a smaller proportion of residents under age 18, a larger proportion of residents with longer commutes, and a higher poverty rate. Note that interpretation of the signs of the variables in the probit regressions is somewhat ambiguous, because the dependent variable is reported damage. A positive sign on any variable could mean that the variable makes damage more likely, or that it increases the likelihood that damage will be reported and entered into the records.

Table 6.10 reports the analysis of the determinants of the log of damage, conditional on a tornado having non-zero reported damage. Note that the natural logarithm of population density, income, path length, and the other demographic control variables are employed in these specifications, and so the coefficients of these variables are elasticities. The results for 1950–2007 as well as separately for 1950–1995 and 1996–2007 are very similar, so we will only discuss the full sample results here. Path length (not surprisingly) increases damage, but the elasticity is around +.4, meaning that a 10% increase in path length increases damage by about 4%, so damage increases less than proportionally with path length. A higher population density also increases damage, and again the elasticity of damage is around +.4, so a 10% increase in density increases expected damage by about 4%. We observe a positive and significant time trend in damage over all samples. As we saw earlier, total damage or damage per tornado has not been increasing over time, and especially since the 1980s. And yet at the level of the individual storm, a tornado in 2007 results in more damage than a comparable tornado in 1950. A part of the time trend could be due to the effect of inflation combined with the intervals for reporting damage. Consider, for example, a tornado with damage in the $50 million–$500 million category. We apply the midpoint as damage for each tornado in this interval and then adjust for inflation, so a tornado that occurred in the 1950s will produce greater real damage than one occurring in the 2000s. We again observe the surprising negative relationship between tornado-path real income and real damage. A 10% increase in income reduces damage by 3%–7%, depending on the specification, but higher-income communities have higher property values and more personal property, which should increase damage. As we will see presently, the relationship between income and damage is complicated, and ceteris paribus income may increase damage.

Damage escalates with the increasing F-scale rating of the tornado. Figure 6.5 illustrates the relationship between reported damage and F-scale, using the 1950–2007 regression model. Damage is scaled so that an F0 tornado is assumed to result in $1,000 in damage, and the figure indicates damage from a tornado on a similar path with a higher F-scale rating. Expected damage escalates quickly—an F1 tornado would produce $7,000 in damage,

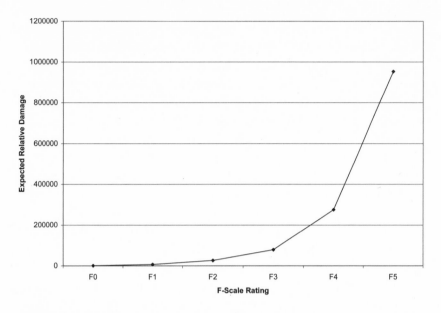

FIGURE 6.5. The effect of F-scale rating on expected tornado damage. Damage amounts in 2007 dollars. Damage from an F0 tornado is normalized to $1,000, and the other amounts represent damage from a comparable tornado with a higher F-scale rating. Based on the 1950–2007 damage regression model in Table 6.10.

an F2 tornado $26,000, an F3 tornado $79,000, an F4 tornado over a quarter million dollars, and an F5 tornado almost $1 million in damage. Recall that stronger tornadoes also tend to have longer path lengths, so a mean comparison of damage by F-scale would be even more pronounced. The damage gradient for F-scale in the 1996–2007 period is even steeper, with an F5 tornado resulting in over 30,000 times the damage of a comparable F0 tornado. As we saw in Table 6.5, although the probability of positive damage essentially plateaus at F2, reported damage does not plateau, and F4 and F5 tornadoes are substantially more devastating than even F3 tornadoes.

The addition of extra demographic variables for tornadoes with damage amounts really alters only the effect of income on damage. As for the probability of damage, income changes sign with the other control variables included, and a 10% increase in income increases damage by about 5%. A larger proportion of mobile homes increases damage, with an elasticity of around +.2. Mobile homes have lower assessed values on average than permanent homes, but their susceptibility to damage ends up increasing total expected damage. Note that mobile homes did not affect the probability of positive damage. An increase in the rural population *decreases* expected damage, which was the opposite of the effect of rural on the likelihood of a damage report. For a given population density, a larger rural population increases the likelihood of some property damage occurring and being reported, but we see that a more urban county is prone to larger damage, given that damage occurs. Counties with older homes experience more damage, as well as counties with a higher proportion of long commutes and a higher poverty rate. We also find that a larger proportion of men and persons over age 65 reduce expected damage. Several of these additional control variables are correlated with income, particularly mobile homes and lower poverty rates. When we include only income as a control variable, income reduces damage, whereas when controlling for these other factors, an increase in income (which might correspond to property values and the amount of personal property) increases damage, as expected.

A number of papers including Brooks and Doswell (2001) have normalized damage to compare losses from weather events over time in ranking the most destructive tornadoes. The normalizations adjust damage from historical events for population and wealth changes as well as for inflation. The idea behind the normalizations is to estimate the damage that might be expected if the same path was struck by a comparable tornado (or hurricane or flood) today, given observed changes in population and wealth since the historical event. Damage is assumed to increase proportionally with population and wealth, so if population of the affected area has doubled since the event, we would expect damage to double as well. Our analysis shows us that tornado damage increases with population, but less than proportionally. The normalization procedure essentially assumes that the coefficients on population and income equal +1.0, but these hypotheses are rejected with our data. Two caveats apply to this, in addition to the general limitation of the precision of our damage estimates. First, for tornadoes at least, county population density is an imperfect proxy for the population along the actual tornado path, and this might bias downward our estimate of the population elasticity of damage. Second, we are using income, which is a flow measure, instead of wealth, which is a stock. While wealth and income are correlated, we see that the relationship between income and damage, at least for tornadoes, is quite complicated. When we compare high- and low-income counties, several other factors related to damage also change, notably the proportion of mobile homes and age of the housing stock. Even when controlling for these other factors, our estimate of the income elasticity of damage, while now positive, is still less than the +1.0 implied in damage normalizations. Consequently we suggest that normalized damage analysis should be attempted with caution.

As discussed earlier in the chapter, some tornadoes appear to simply be missing damage reports, since given their path length, F-scale rating, and casualties it seems impossible that the tornado did not produce any property damage. Consequently we concluded that the zero damage category mixes together tornadoes with no damage and those with positive but missing damage information. A dummy variable that merely controls for a positive damage report could mix these reasons for no report. To investigate if missing data were affecting our analysis, we ran the models reported in Tables 6.9 and 6.10 excluding those tornadoes with seemingly missing damage reports. Specifically, we considered any tornado rated F2 or stronger with a zero damage report to be anomalous and excluded these storms. Surprisingly, exclusion of these tornadoes had almost no effect on the estimates of either the probit or damage models over any of the time periods, and consequently we do not present these results.[2]

6.6. Community Impacts of Tornadoes

Natural disasters have the potential to severely affect communities, beyond damaging or destroying individual homes and businesses. Community-wide effects result from damage to infrastructure, disruption of the local economy, and the coordination problem inherent in rebuilding. The ultimate impact of a disaster on a community depends on community size, the type of economic base, and residents' commitment to the area. The Galveston hurricane of 1900, the 1906 San Francisco earthquake, and Hurricane Katrina in 2005 illustrate the potential for a disaster to affect the long-run growth trajectory of a city; Smith et al. (2006) demonstrated that by the year 2000, some neighborhoods in the Miami area had yet to fully recover from 1992's Hurricane Andrew. Overall, however, the long-term experience of most areas hit by disaster is one of resilience. This fact was observed as far back as 1848 by John Stuart Mill, when he noted ". . . the great rapidity with which countries recover from a state of devastation; the disappearance, in a short time, of all traces of the mischiefs done by earthquakes, floods, hurricanes, and the ravages of war . . . and yet in a few years after, everything is much as it was before" (Mill 1848).

The overall impact of a tornado on a local economy will depend on the preparedness of the business community. When businesses fail to adequately prepare or insure, they can be closed longer than necessary or may eventually fail; one-quarter of businesses that close after a disaster never reopen (Institute for Business and Home Safety). Research has found several factors that consistently affect business preparation. One is size, with large businesses more likely to prepare than small businesses (Webb et al. 2000). Businesses that own their facility are also more likely to prepare than those that lease (Tierney 1997; Webb et al. 2000), which suggests that when building owners and business owners are separate, they have difficulty coordinating for hazards preparation. A business's perception of hazard vulnerability affects preparations, but the relationship is complex. For example, although previous hazard experience does not lead to more preparations, a previous hazard *disruption* does increase preparation (Webb et al. 2000). Perceptions of risk drive preparations by small businesses, but unfortunately perceptions often fail to correlate with objective measures of risk such as location in a flood plain (Yoshida and Deyle 2005).

A number of factors affect whether businesses close and the duration of the closure. Disruption of the transportation network in the Northridge earthquake affected employees, suppliers, and customers and contributed to many closures (Gordon et al. 1995). Businesses also require time to clean up, even if damage to the facility is modest (Tierney 1997). However, utility lifeline disruptions are typically more significant than actual damage to a business's facility. Only 15% of businesses surveyed after the 1993 Des Moines flood actually had flood damage, but 80% lost water service and 40% sewer service, and it was these utility disruptions as opposed to the flood waters that led to most business closures (Webb et al. 2000).[3] Tornadoes are unlikely to inflict the type of widespread damage to utility networks commonly experienced with hurricanes, floods, ice storms, and other weather hazards.

Overall research shows that businesses often recover from disasters, and larger firms are more likely to recover (Webb et al. 2000). Several research findings are negative, regarding factors that do not seem to influence recovery—notably, previous disaster experience, the extent of predisaster preparations, and the use of external resources like loans from the Small Business Administration (Webb et al. 2003). Research also finds a relatively low level of preparation for hazards by businesses. For example, Webb et al. (2000, p. 84) conclude, "The average or typical business places relatively little emphasis on disaster preparedness and other loss reduction mechanisms."

The lack of establishment-level preparations may stem from the role of utility disruptions in driving closures. The utility network is a public good for the business community, and preparations to enable a business to quickly reopen will be wasted if utility disruptions prevent prompt reopening. On the other hand, if utility outages force a business to close after a disaster, firms have plenty of time to make repairs even if they have not prepared extensively before the event.

Economic effects of weather events depend greatly on the size of the storm. Hurricanes can impact an entire region, and several studies have focused on economic recovery from hurricanes, including Hurricane Hugo in South Carolina in 1989 (Guimares, Hefner, and Woodward 1993) and Hurricane Andrew in 1992 (West and Lenze 1994). Belasen and Polachek (2009) found that stronger hurricanes in Florida have a greater impact on employment than weaker storms.[4] A second factor affecting recovery is the health of the local economy prior to the event. Swift recovery is more likely if a community had a diversified, thriving economy; on the other hand, if the community was fragile before the storm, the event may be a tipping point.

Tornadoes are much smaller events than hurricanes, although a large tornado outbreak can result in damage across a region. Nonetheless, tornadoes can devastate small towns and occur throughout much of the nation. In April 2007 a tornado struck the small town of Tulia, Texas, heavily damaging the town's business district. The town's only grocery store did not reopen after the twister, and residents had to drive 30 miles to Amarillo for grocery shopping (Martinez and Ewing 2008). Picher, Oklahoma, an EPA Superfund site, was struck by an F3 tornado in 2008. Even though the town was designated as a Superfund site and all residents were made offers on their property to leave, some had stubbornly chosen to stay. But the tornado changed that. Over 150 homes on the south side of town were destroyed, which proved to be the final nail in the coffin of this once-thriving mining community. The school district graduated its final class in May 2009, the Post Office closed in July, and the municipality ceased to operate in September of that year (Stogsdill 2009).

Labor and housing markets provide early clues regarding a community's prospects for recovery. If the economy has been permanently slowed, the effects should be revealed first in these two markets. Windstorms can directly and indirectly affect the residential housing market—most homes are not engineered structures and thus are vulnerable to wind damage and damage from wind-blown debris, directly reducing the supply of housing. Another direct effect would be a possible demand surge if reconstruction brings a substantial number of new workers to the local area. In the longer term, storm damage can highlight vulnerabilities in standard construction practices and thus create a demand for better-built homes and eventually stricter building codes. Improved construction will typically increase costs, resulting in an increase in home prices. Windstorms can also indirectly affect housing markets through the expected future health of the area. Demand for housing is contingent on the overall economic health and expected future health of the community; if major businesses decide to relocate following a disaster, housing prices will fall and the expectation of excess supply into the future will reduce new construction, beyond reports to existing homes.

Ewing, Kruse, and Wang (2007) examined the price indices for housing in six metropolitan statistical areas (MSAs) affected by windstorms—tornadoes and hurricanes. Markets affected by tornadoes were Nashville (1998), Oklahoma City (1999), and Fort Worth/Arlington (2000), while the hurricane-affected markets were Corpus Christi (Hurricane Bret 1999), Miami (Hurricane Andrew 1992), and Wilmington, North Carolina (Hurricanes Bertha and Fran 1996 and Bonnie in 1998). Tornadoes caused a decline in the price index of between .4% and 1.8%. Hurricanes resulted in price declines similar to tornado-affected markets, by the 4th quarter after the event, the declines had ceased. Ultimately, the recovery of the housing market depends on the overall health of the regional economy.

Labor markets also reflect the health of the regional economy, and a tornado can alter the demand for labor by damaging the general economy or permanently damaging infrastructure necessary to the region. Tornadoes can cause a reduction in the level of employment (e.g., an increase in unemployment) and an increase in employment volatility, both of which are harmful. In 1999, a large tornado outbreak affected central Oklahoma and southern Kansas; the following year, a series of tornadoes tore through Fort Worth and Arlington, Texas. These windstorms affected urban areas with large and diverse populations. Studies of the labor market reaction to the Oklahoma and Fort Worth tornadoes provide insights on the effect of tornadoes on large local economies (Ewing, Kruse, and Thompson 2003, 2005, 2007). Results suggest that the 1999 Oklahoma tornadoes had a negligible effect on the labor market in central Oklahoma and southern Kansas; only one sector, government, showed an adverse effect. The results in Fort Worth are similar in showing a reduction in growth in the labor market, but not a contraction in overall employment. The results further suggest that labor market volatility decreased after the tornadoes. Thus, although tornadoes represent a serious threat to small economies, a large metropolitan economy is usually quite resilient to these storms.

6.7. Conclusion

Tornadoes cause destruction to property and can impact businesses and communities. In some cases a tornado can even provide a death blow to a struggling small community, as the case of Picher, Oklahoma, illustrates. Generally, however, the impacts of tornadoes on a local economy of even modest size are quite limited, as damage paths are narrow and there tends to be little disruption to utility networks. Damage from tornadoes has been less studied than casualties. But this presents interesting research opportunities since damage is used in casualties analysis as a measure of impact from a tornado on buildings. However, the old cautionary line from computer science—"Garbage In, Garbage Out"—must be kept in mind here, because the quality of damage reports is not high. Although differences in casualties per million dollars of damage exist across the United States, they seem to coincide with the Southeastern component to mobile-home, nocturnal, and off-season vulnerability already identified in Chapter 3. The overall level of damage is modest; the greatest burden of tornado damage is in Kansas and Oklahoma, where damage averages an estimated $39 per person per year. Perhaps the most notable result from our analysis of damage is the rather complicated relationship between income and damage. Intuition suggests that damage should increase with income, as wealthier communities have more real and personal property at risk. But other economic factors that correlate with income appear to obscure this relationship, and overall it appears that a given tornado imposes a greater absolute toll (and even more so in relative terms) on lower-income communities.

6.8. Summary

While fatalities from tornadoes typically grab the headlines, tornadoes cause a great deal of property damage as well. As is the case with casualties, a small proportion of tornadoes causes the majority of property damage. In contrast with hurricanes, damage from tornadoes has not been increasing over time. Not surprisingly, the F-scale rating of the tornado and the damage-path length are two important determinants of damage. Damage exhibits a Southeastern bias as was observed with casualties, although per capita damage is greatest in Kansas and Oklahoma.

6.9. Appendix

We employ a joint analysis of the determinants of the probability of a tornado having a positive damage report and the amount of damage if it occurs. A simple regression analysis of damage would consequently suffer from selection bias, since tornadoes that have positive damage are not a randomly selected subset of all tornadoes. Instead, we employ a two-stage sample selection model to control for non-random selection (Heckman 1976; Greene 2000). The first stage is a probit regression of whether a tornado had positive damage; the second stage is an OLS regression of the amount of damage only for tornadoes with damage. Inclusion of the inverse Mills' ratio (IMR) in the second-stage OLS regression controls for the non-random selection of tornadoes with a given type of damage. The IMR equals the ratio of the probability density and cumulative distribution functions of the normal distribution, evaluated at the estimated value of the probit index function for a tornado. A sample selection model allows the control variables to have different impacts on whether a tornado had positive damage as well as the amount of damage. For instance, our sample selection model indicates that a larger rural population increases the probability that a tornado will result in positive damage, and yet reduces the amount of damage when it occurs.

7

**GOING FORWARD: USING SOCIETAL IMPACTS
RESEARCH TO REDUCE TORNADO RISK**

7.1. Introduction

In this book we have examined how tornadoes produce fatalities and injuries, how tornado warnings provided by the NWS affect casualties, how
impregnable shelters can be built to prevent casualties, and examined patterns in property damage. We conclude in this chapter by bringing together
the pieces of our analysis to estimate the overall societal impact of tornadoes
and to examine how NWS warnings have affected and continue to affect
this impact. We then offer some suggestions based on the vulnerabilities we
have identified to reduce tornado-related casualties in the future. Normally,
science proceeds by its own internal logic, with discoveries producing new
puzzles but also pointing the way forward. Societal impacts research enters
as an addendum, to try to determine the value to society of the next question
that the scientific process will answer. Here we offer suggestions based on
societal impacts for future directions in research. We conclude by considering whether the United States is approaching the optimal number of tornado
fatalities that can be expected, and offering some lessons from our study for
research on the societal impacts of weather generally.

7.2. Assessing the Societal Cost of Tornadoes

We can now pull together the components of our analysis to assess the over-all impact of tornadoes in the United States. We provide a monetary value of the various impacts, using the values of time and statistical lives and in-juries discussed in Chapter 4. The cost of tornadoes has three components: casualties, property damage and other indirect economic impacts, and the response to warnings. Casualties and property damage are obvious impacts, while indirect economic impacts like the specific value of business closings and utility disruptions are less visible, and will not be assessed here due to a lack of such studies focusing on tornadoes. The cost of responding to warnings is even less immediately obvious, but it is nonetheless a real cost, as people must interrupt their daily activities to shelter.

We make our assessment using impacts from 1996 to 2007. Doppler radar installation was largely complete by 1996, so casualty totals will reflect radar's life-saving effects; in addition, 1996 was also the year that damage estimates were employed in place of damage intervals. Moreover, the increase in the number of tornado warnings issued annually by the NWS due to the Doppler network had already occurred by 1996. We have discussed the derivation source of damage estimates in earlier chapters, so we will just summarize the results here. Table 7.1 presents the totals. Between 1996 and 2007, annual inflation-adjusted property damage, fatalities, and injuries averaged $1.123 billion, 63.3 persons, and 999 persons, respectively.[1] Time under tornado warnings averaged 234 million person hours between 1996 and 2004, and we estimated that half of this time, or 117 million person hours per year, might be spent sheltering. To yield the dollar totals in Table 7.1, we applied the monetary valuations discussed in Chapter 4: that is, $7.6 million per fatality; injuries at 1/100 of this value, or $76,000; and time at $11.38 per hour. We calculate the total the cost of tornadoes in the United States in two ways—in column A using time under warnings as the cost of tornado warnings, and in column B using time spent sheltering as the cost of warnings. The percentage of the total cost that each component contributes is also reported in brackets for each calculation. The total cost of tornadoes is larger using time under warnings: $4.3 billion per year versus $3.0 billion using time spent shelter-ing. In either case, tornado warnings represent the largest component of the cost of tornadoes. The monetary cost of fatalities is $481 million and injuries $76 million, so property damage is about double the cost of casualties. These totals are subjective in accordance with the caveats discussed in previous chapters, like the limitation of damage estimates and paucity of evidence on

TABLE 7.1. The Annual Impact of Tornadoes

Impact	Average Annual	Monetary Value (A)	Monetary Value (B)
Property Damage	$1.123 billion	$1123 [25.8%]	$1123 [37.3]
Fatalities	63.3 persons	$481 [11.1%]	$481 [16.0%]
Injuries	999 persons	$76 [1.7%]	$76 [2.5%]
Time Under Warnings	234 million person hours	$2670 [61.4%]	
Time Spent Sheltering	117 million person hours		$1334 [44.3%]
Total		$4348 [100.0%]	$3014 [100.0%]

Damage and casualties are averages for 1996–2007; time under warnings is an average for 1996–2004. Monetary value is calculated in column A using time under warnings as the cost of tornado warnings, and in column B using time spent sheltering. The valuation of lives lost, injuries, and time under warnings is discussed in the text.

the response rate for tornado warnings. Our totals also exclude any indirect losses from business closures or utility disruptions.

The contribution of tornado warnings to the social cost of tornadoes may surprise many readers, and results from two factors: the installation of Doppler weather radars in the 1990s, which increased the number of warnings issued annually, and the success of NWS efforts to reduce the lethality of tornadoes (Doswell et al. 1999). Table 7.2 presents two counterfactuals for the cost of tornadoes to help us see how tornado warnings affect the social cost. Column A repeats the cost figures with time spent sheltering from Table 7.1. The first case, as reported in column B, considers how costs might look today if not for tornado warnings. We retain the current level of property damage, but casualties would be greater and the cost to responding to warnings would be eliminated. To estimate a possible casualty rate, Brooks and Doswell (2002) estimated that the smoothed, 25-year U.S. tornado fatality rate declined from 1.8 per million in 1925 to .11 per million in 2000, as efforts to reduce tornado lethality really began following the devastating Tri-State Tornado in 1925. Similarly, our time series of injuries in Chapter 3 indicates that the smoothed injury rate has fallen by 75% over the same period. If these reductions in casualty rates had not occurred, the United States today would experience an average of 1,036 fatalities and 4,000 injuries per year, instead of the recent averages of 63 and 999, respectively. The monetary cost of tornadoes in this case, as reported in column B of Table 7.2, would be $9.3 billion annually, with $7.9 billion stemming from fatalities and $300 million from injuries. To place the hypothetical fatality total in this scenario

TABLE 7.2. The Societal Cost of Tornadoes and the Role of Warnings

	A	B	C
Damage	$1123	$1123	$1123
Fatalities	$481	$7876	$481
Injuries	$76	$304	$76
Time Sheltering	$1334		$586
Total	$3014	$9303	$2267

Amounts are in millions of 2007 dollars. Column A represents the current impacts of tornadoes from Table 7.1. Column B reflects higher casualty rates but no time spent sheltering from a counterfactual with no tornado warnings. Column C reflects the lower sheltering costs with SBWs for tornadoes.

in perspective, it is approximately equal to the direct death toll in Hurricane Katrina, and it would be occurring every year. Comparison of columns A and B of Table 7.2 shows that tornado warnings reduced the cost of tornadoes by about two-thirds, even though response to warnings has now become the largest component of the $3 billion cost per year. (In reality, it is probably excessive to attribute the entire reduction in casualties to the efforts of the NWS, but the totals do reflect what could happen if the lethality of tornadoes was unchanged since the 1920s.)

Column C of Table 7.2 reports projected costs of tornadoes with Storm Based Warnings (SBWs). These new warnings were implemented for tornadoes and other types of severe storms nationwide in October 2007, and have the potential to reduce the area of the typical tornado warning by 70%–75%, as discussed in Section 4.8. As argued there, time spent under warnings should be reduced less than this amount, because residents of large counties who are far from a possible tornado probably are not sheltering currently. Nonetheless, we estimated that SBWs would reduce the time spent sheltering by more than 50%, or $747 million per year.[2] With this change, the total cost of tornadoes will be reduced by 25% over the current level to $2.3 billion per year, and property damage will now be the largest component of cost.

7.3. The Determinants of Casualties: Keys for Future Reduction

Our analysis of tornado casualties illustrates some sources of high vulnerability and consequently provides directions for future efforts to reduce casualties. The differences in casualty rates across the day or year represent theoretically feasible reductions in casualties, since society already achieves lower casualties for tornadoes in the early afternoon or in the summer. Thus the potential should exist to address some of the sources of high vulnerability. We do not offer a specific plan for realizing these reductions, but do quantify the percentage reduction in overall casualties possible in each case. Note that the reductions in overall casualties are all partial equilibrium effects, or the reduction possible from the current level of casualties if only this source of vulnerability is reduced. Different reductions will apply if other causes of casualties have already been addressed.

Mobile-home fatalities. The rate of fatalities in mobile homes exceeds the rate for permanent homes by at least a factor of 10. Tornado fatalities in the United States could be substantially reduced if the mobile-home fatality rate could be brought in line with the permanent-home rate. To see how much reduction is possible, suppose that the mobile-home fatality rate could be reduced by 90%. The U.S. averaged 29.25 mobile-home fatalities annually between 1996 and 2007, so a 90% reduction in the mobile-home fatality rate would save almost 27 lives per year and reduce the current national fatality rate by 43%.

To accomplish this goal, residents of mobile homes need to have safer places to shelter. Analysis shows that mobile-home fatalities occur disproportionately in weaker tornadoes, and that F3 tornadoes specifically account for much of this death toll. Permanent-home fatalities are rare in F3 and weaker tornadoes, which suggests that permanent homes typically provide enough protection for residents to avoid fatal injury except in the strongest tornadoes. Tornado shelters represent one way to protect mobile-home residents, and offer cost-effective protection based on comparisons with the value of a statistical life observed in the market, at least in the most tornado-prone states. Community shelters have even lower costs and could be provided by mobile-home parks. Indeed, many parks already provide shelters as an amenity for residents; Schmidlin, Hammer, and Knabe (2001) report that 80% of parks in Kansas surveyed did in fact offer shelter, while the Oklahoma Manufactured Housing Association (2005) reports that about 60% of parks in Oklahoma offer shelters. Shelters have a history of protecting residents of manufactured homes. The 1991 Wichita, Kansas F5 tornado killed 13 residents and destroyed 233 of 241 units at a mobile-home park in Andover, Kansas, but 200 residents survived in the park's underground shelter (Grazulis 1993). Parks have the ability to supply residents of manufactured homes with other tornado safety services as well, such as tornado sirens. However, the efforts of parks cannot fully address the mobile-home vulnerability, as more than half of these housing units nationally are not located in parks.

Another approach would be to make manufactured homes more wind resistant. The Department of Housing and Urban Development (HUD) added wind load provisions to the HUD Manufactured Housing Code in 1994. The wind load requirements were intended to address the vulnerability of these structures to hurricane winds (De Alessi 1996); 98% of mobile homes in the path of Hurricane Andrew in 1992 were destroyed, as compared with 11% of permanent homes (Grosskopf 2005; Rappaport 2000). The wind load provisions do appear to have reduced the damage caused by hurricanes, as revealed by the hurricanes that struck Florida in 2004–2005 (Grosskopf 2005), and they seem to be reducing the vulnerability of mobile homes to tornadoes as well. In the tornadoes that struck Lake County, Florida, in February 2007, only 9% of mobile homes installed after the 1994 wind load provisions were enacted were leveled, compared with 24% of homes installed prior to the implementation of the HUD code for manufactured housing in 1976 (Simmons and Sutter 2008). The reduction might seem modest, but fatalities occurred primarily in leveled homes (16 of 17 fatalities for which we could document the condition of the damaged structure), and no fatalities occurred in any home installed after 1994. These results suggest that construction adhering to these wind load provisions might reduce mobile-home fatalities by up to 70% relative to pre-1976 homes.

The continuing vulnerability of mobile homes stands in contrast to the safety record of businesses, where only 5% of tornado fatalities occur. While tragic cases exist, such as the 1993 tornado in Colonial Heights, Virginia, that killed two cashiers and a patron at a Walmart Super Center, anecdotes of customers and employees saved by prompt actions and preparations abound. To evaluate the vulnerability of businesses, we would need to know what proportion of persons in tornado paths were at a business location. Given the frequency of tornadoes in the afternoon and early evening, when people are likely to be at work or shopping or dining, it seems likely that more than 5% of persons in the paths of tornadoes would be at business locations, and if this is indeed the case, we could conclude that businesses are relatively safe locations. Of course, mobile-home parks are a type of a business, too, and yet we do not always see the safety effects of businesses apply in these locations. Particularly troubling in this regard is the low percentage of parks in the Southeast providing shelters (Schmidlin, Hammer, and Knabe 2001), as 58% of mobile-home fatalities occur in the Southeast. In addition, manufacturers should have an interest in improving their product if possible and making it cost-effective, because as discussed in Chapter 5, tornado risk induces some residents to choose relatively safer site-built homes.

Tornadoes after dark. Tornadoes that occur during the late evening or overnight result in substantially more casualties than comparable storms during the afternoon. If the lethality of nighttime tornadoes (midnight to 6 AM) could be reduced to afternoon levels, expected fatalities in these tornadoes could be reduced by 68% and expected injuries by 32%, based on the 1986–2007 regressions from Chapter 3. Over these years, 15% of fatalities and 13% of injuries occurred during the overnight hours, so with this reduction in the lethality of nighttime tornadoes, overall fatalities could be reduced by 9% and expected injuries by 4%. Moreover, if casualties from tornadoes at all times of the day could be reduced to the lowest levels observed, fatalities could be reduced by 31% and injuries by 15%.

Obtaining a reduction in nighttime lethality would probably require several components. Part of the nighttime vulnerability also likely stems from less effective warning transmission. The improvement in warning skill since the modernization of the NWS and the refining of warning areas with SBWs increases the value of warnings—and by extension the value of NOAA Weather Radios or other alert systems to deliver warnings when residents are asleep—but residents must own a Weather Radio to receive the warning. Nighttime fatalities occur disproportionately in mobile homes, and thus nocturnal vulnerability is intertwined with the mobile-home problem. Residents of mobile homes require a sheltering option they will actually undertake, as Schmidlin et al. (2009) found that about 70% of these residents did not respond to tornado warnings. Residents just do not seem willing to abandon their homes for a ditch during the night, and the near certainty of discomfort from being outside in the rain might well outweigh the still-low probability of a tornado actually striking their home.

Off-season lethality. We have documented the substantial influence of the month of the year on fatalities and injuries. The most powerful tornadoes occur during April and May, and yet tornadoes in the winter months are much more lethal, everything else being equal. If tornadoes in February or November were only as dangerous as comparable tornadoes in July or June, casualties could fall substantially. To quantify the possible reduction in overall casualties, we need to specify in which months casualties could be reduced and to what level. In the conservative case that casualties could be reduced in only the most lethal off-season months of February, October, November, and December, and then to only a moderate level (the May fatality rate and July injury rate), fatalities could still be reduced by 11% and injuries by 8%, based on the reduction in casualties possible in each month and the proportion of fatalities occurring in each month. In the limiting case where casualties in each month could be reduced to the level of the least dangerous month (July for fatalities and May for injuries), the nation could experience a 72% reduction in fatalities and 36% reduction in injuries. As discussed in Chapter 3, the "month of the year" effect is probably the least intuitive of the vulnerabilities we have identified. The days are shorter in the winter months, which means that the potential exists for more tornadoes to occur after dark and for this vulnerability to be simply an extension of the nocturnal vulnerability. However, the month effect is larger in amplitude than the day-part effect, with only a portion of tornadoes likely to be occurring after dark compared to spring or summer months. We found relatively minor differences in warning performance across the months, and in any event the difference in casualties between warned and unwarned tornadoes is smaller than the month effect. Future societal impacts research is clearly warranted in this area to determine what drives the very large difference in casualty rates across the year.

Improved warnings. The quality of tornado warnings has improved markedly since the 1980s, as illustrated in Figures 4.5 and 4.6. But warnings could still be improved, further reducing tornado casualties. To estimate the magnitude of potential gains from further improvements in warnings, we calculate the change in casualties that would result if all "underwarned" tornadoes were optimally warned. The lowest fatalities occur in the 6- to 10-minute lead time interval, and the lowest injury rate in the 11- to 15-minute range. Thus, we consider tornadoes that were unwarned or that had a lead time in the 0- to 5-minute range as underwarned. Using the distribution of lead times over the years 2000 to 2004, 36% and 10% of tornadoes occurred in these two categories, 6 to 10, and 11 to 15 minutes. With optimal warnings, fatalities could be reduced for unwarned tornadoes by 52%, and for tornadoes with lead times of 0 to 5 minutes by 40%, while injuries in these intervals could be reduced by 42% and 12%, respectively. Tornadoes in these categories accounted for 19% and 5% of fatalities and 29% and 5% of injuries, respectively, so applying the potential reductions in casualties to the percentage of casualties in tornadoes of these types shows that optimized warnings could reduce fatalities by another 12% and injuries by 13%.

7.4. Societal Impacts as a Guide for Meteorological Research

We begin this section with a frightening scenario. Imagine a tornado, similar to the 1925 Tri-State Tornado in strength, path length, and destructive power, already producing F5 damage and showing no signs of weakening. Now imagine that it's approaching a major metropolitan area. This is the extreme urban impact tornado explored by Wurman et al. (2007), and their analysis shows that hundreds of thousands of persons could be in the path of such a monster tornado. Our analysis of tornado shelters reveals that permanent-home fatalities are simply too infrequent to make shelters cost-effective, and hardening targets to survive an F5 tornado is not something that's going to happen. As economists, we could point out that this scenario, as tragic as it might be, is a low enough probability event that society would rationally choose not to mitigate for it in advance. But if hardening all buildings is not feasible, are there any alternatives that might reduce fatalities in this event?

One strategy in this case would be for residents to get out of the path of the tornado. Indeed, some people already "evacuate" for tornadoes; Hammer and Schmidlin (2002) found that 47% of residents surveyed who had been in the path of the 1999 Oklahoma City F5 tornado fled the storm. None of these "evacuees" was injured or killed, and half fled in a vehicle. Hammer and Schmidlin note that "[m]any people recognized that it was safer to be in any location outside of the tornado path" and contend that this shows that "people can successfully make reasoned decisions that reduce the risk of injury and death" (p. 580). Residents today would most likely be aware of an approaching super tornado, and no action short of getting out of the storm's path would prevent large-scale loss of life.

It's easy to formulate half a dozen potential serious problems with tornado evacuations. We do not deny the problems, but instead wish to consider the existing problems as outlining the research program required to make evacuations a viable, and in our minds a low-cost, strategy to deal with long-track, violent tornadoes. Consider the weather and warning information needed to make evacuation work. With perfect information we would know the exact path the tornado will take, when it will hit each point on the path, and how long it will maintain F5 strength. We could identify the homes and businesses in the tornado path that lack a shelter or safe room capable of withstanding F5 winds and therefore need to be evacuated. We would also know which homes were safely out of the way and do not need to be evacuated, reducing the potential for extra evacuees to create congestion and interfere with timely evacuation, or to inadvertently move into the tornado's path. Finally, as the tornado is already on the ground, there would be no potential for an unwarned event or a false alarm. The distance residents must travel to get out of the damage path is modest—the widest tornado on record is about two miles, and paths even one mile wide are very rare. Consequently, residents would only have to go perhaps half a mile to get out of harm's way, and may well have enough time to evacuate on foot. With sufficient lead time, friends, family, and neighbors may be able to assist mobility-impaired residents (disabled persons and households without cars) in getting out of the path of the tornado. But of course, all of these scenarios would require perfect information, which we do not have. There would also be a serious communications problem in disseminating the warning and evacuation message in a timely fashion and in a way that residents can understand, particularly if evacuations are only encouraged for a handful of the most powerful tornadoes. Clearly, accurate information can make a big difference. As Hammer and Schmidlin's study demonstrates, on May 3, 1999, Oklahoma residents in harm's way received information that allowed them to successfully get out of the way of this tornado.

Perfection is impossible, and the problems with evacuation are likely to escalate quickly with each minor error in the tornado-path forecast. Even a one-mile band of uncertainty on either side of a forecast path would triple the area that might evacuate. Congestion could mount, evacuees could fail to get out of the path, or people could even evacuate into the path of the storm and face greater risk outdoors or in a vehicle than in a permanent home. However, due to the short distance evacuees need to travel to get out of the path, it's possible that evacuations could be staggered, keeping 10 miles in front of the storm, for example.

From our vantage point, meteorologists appear to be making progress toward making at least selective evacuations effective. Increasing numbers of storm trackers allow quick recognition of tornadoes on the ground and that possible F5 damage is occurring. NWS Doppler radars have improved tornado warnings, but currently do not observe the lowest portion of thunderstorms and thus the actual tornado circulation and wind speeds. However, mobile Doppler radars are increasingly being deployed and are able to observe the low-level winds in a tornado, and future generations of weather radars may be able to provide even better observation of the low levels of thunderstorms where the tornadoes are (see Golden and Adams 2000; National Research Council 2002). Currently, forecasts of the path of a tornado are not sufficiently accurate for evacuations (consider the large size of the warning polygons for SBWs), but we suspect that as VORTEX II and mobile Doppler radars allow meteorologists to observe low-level circulation in supercell thunderstorms on a regular basis, science's understanding of tornadoes will improve rapidly.

We can offer an estimate of the value to society of being able to implement evacuations for the strongest tornadoes. Essentially, the value of the research (if successful) would be to avoid the type of disaster described by Wurman et al. (2007). Based on our evaluation of the worst-case scenario using the fatalities regression models, we predicted a worst-case fatality total of 2,312 in Section 3.8, while tornadoes in several other counties were capable of producing a fatality total over 1,000. Let's round off and say that a worst-case death total of 1,000 could be avoided if residents could evacuate for this hypothetical long-track tornado.[3] So how likely is it that such a tornado will hit a major metropolitan area? We are taking a guess here, but for the purposes of illustration, let's say that this is a one in 500-year event. There were several tornadoes during the 20th century that may have been capable of such devastation had their path had been through a major metropolitan area. In this case, then, a policy of evacuation would save two lives per year over 500 years, and at a value of $7.6 million per statistical life, evacuations for an urban F5 tornado would produce expected benefits of about $15 million a year. If a sufficient understanding of F5 tornadoes of an urban was available, evacuation might also allow life-saving evacuations for other F5 tornadoes. For instance, residents of Greensburg, Kansas, could have gotten out of town prior to the May 4, 2007, F5 tornado.

Strong and especially violent tornadoes drive the casualties and damage from tornadoes. Casualties are also concentrated on days of large tornado outbreaks: a mere 10 days have accounted for 30% of fatalities since 1950. Violent tornadoes are particularly likely to occur in these large tornado outbreaks: 603 of 606 F4 and F5 tornadoes between the years of 1950 to 2007 occurred on days with at least one other tornado, and an average of 19 tornadoes occurred on the days with a violent tornado. Since in terms of societal impacts all tornadoes are not created equal, neither should the warning process treat all tornado risk equally, so in light of the concentration of societal impacts, tornado warnings, watches, and convective outlooks could be directed more toward alerting people to violent tornadoes or days on which major tornado outbreaks are possible. Residents will benefit from advance knowledge of tornado strength: The value of sheltering upon receipt of a warning depends on whether the resulting tornado is an F0 or F4. Society's willingness to postpone activities that might result in heightened danger (e.g., large outdoor events like concerts or sporting events) will depend on whether a couple of weak tornadoes are possible, or a major outbreak with numerous strong and violent tornadoes is possible. Progress is being made on this front. Vescio and Thompson (2001) discuss an experimental program in 1997–1998 in which forecasters produced subjective probability estimates for one or more and three or more tornadoes occurring within a tornado watch, and the maximum intensity of a tornado within the watch. The subjective forecasts exhibited skill, although the maximum tornado damage potential tended to be overestimated.

The potential for watches and convective outlooks to reduce casualties by identifying major outbreaks with violent tornadoes must be considered with caution, as our analysis in Chapter 4 found that watches had no apparent effect on casualties. As a consequence, it is difficult for us to offer a potential value for research that leads to advances in watches or convective outlooks. Note that the lack of effect of long lead–time warnings or watches might be a consequence of traditional types of tornado precautions. The value of a forecast or warning depends on losses avoided by the action taken based on the warning, so a tornado watch may be of no value when residents fail to shelter appropriately when a tornado approaches. However, if watches begin to alert residents to large outbreaks with multiple violent tornadoes and have sufficient skill, we may see schools dismissed early or sporting events postponed in response. Our inability to document a value of existing tornado watches today does not mean that major outbreak watches would not prove valuable.

The final direction for future research discussed here is the effort to determine if areas of localized tornado risk—tornado tracks—exist and can be identified. Identification of tornado tracks creates value to society because high-vulnerability structures, including manufactured homes, hospitals, nursing homes, and schools, could then be located outside of the tracks. Currently, meteorologists do not believe that well-defined tracks exist. Public perceptions that some towns are "always" or "never" hit by tornadoes are likely incorrect inferences based on a short historical record and/or selective memory. Our inability thus far to identify tornado tracks, however, does not prove that they do not exist. It's quite possible that local terrain may indeed affect the development and movement of tornadoes. Historical observations on tornado paths have never been detailed enough to allow a careful test for differences in where tornadoes have tracked, and radars do not allow observation of the lowest levels of severe thunderstorms. In the future, new radars will observe the lowest levels of storms and it will be possible to map exact tornado paths. Combining the path information with observations of the storm and satellite maps of the terrain could lead to greatly improved understanding of tornado development and movement. Society will benefit even if tornado tracks are merely probabilistic, that is, areas that have a non-zero probability of tornado damage.

Identification of tornado tracks, if they do exist, would potentially have substantial value. To see this, suppose that in a given county the annual probability of tornado damage in an area is currently estimated to be $4*10^{-4}$, and assumed to be uniform across the county. But now suppose that research allows us to refine this estimate, and in fact half the county turns out to face an annual damage probability of $6*10^{-4}$ and the other half an annual probability of $2*10^{-4}$. Over time, it would be possible for mobile homes, nursing homes, hospitals, schools, and sports stadiums to be (re)located in the relatively low-risk part of the county, which would reduce expected fatalities in the county by half. To quickly estimate a value for tornado tracks we limit our attention to mobile-home fatalities. An average of 29.25 mobile-home fatalities occur each year. If tornado tracks exist throughout the country (or at least in states where fatalities regularly occur) where the probability of a tornado is half the current estimated probability, mobile-home fatalities could be reduced by 50% in the long run. This would save 14.63 lives per year, and with the application of the $7.6 million value of a statistical life, the ability to map out tornado tracks would be worth $111 million per year. And this is just from mobile-home casualties—many other vulnerable facilities like schools, nursing homes, and utility infrastructure could also be located in low-risk areas, producing additional benefits.

7.5. Summary

We began our work on tornadoes very much aware that the subject was outside our normal arena and often cautious about the contributions we could offer. But economic analysis is often helpful in identifying hidden costs or benefits. We have seen one prominent example in our analysis, the cost of tornado warnings. Tornado warnings disrupt normal activities, and even though the cost per person is small, with the NWS issuing over 3,500 warnings a year, small disruptions really add up, so much so that until recently the cost of responding to warnings was the greatest single source of societal cost. We doubt that many meteorologists would have recognized this, or that the introduction of SBW for tornadoes would increase the value of tornado warnings by more than reducing the false alarm ratio to zero. What are the other substantial but largely invisible costs of extreme weather?

As we conclude this book we hope that we have highlighted results from our research that can help further our understanding of how society can minimize the impacts of nature's most powerful storm. We also hope that our readers may realize that by allowing a couple of economists to "trespass" on their research sandbox, insights can be obtained that otherwise would have proven elusive.

7.6. Conclusion

Can all tornado fatalities and injuries be eliminated? Seemingly, this might be possible. Safe rooms could be built to protect people from the most powerful tornadoes, and the probability of detection increased to essentially 1.0. Although eliminating tornado fatalities is a noble goal in principle, it is not a sound goal for policy. Tornado fatalities are tragic for the affected families and a cost to society, but reducing tornado fatalities is also costly. People value safety, but they also value other things in life, and reducing tornado fatalities requires resources that could be used toward other goals. The law of diminishing returns implies that it will become increasingly difficult—and therefore costly—to further reduce tornado fatalities, and at some point the remaining casualties will be simply too costly to attempt to eliminate. We have seen this in the high cost per life saved for tornado shelters in some states, and the cost of responding to traditional, county-based tornado warnings. Economics shows us that there will be an optimal number of tornado fatalities, and that this number will not be zero. However, we do not think we have reached the optimal number yet, and have offered some suggestions for reducing fatalities and other impacts in this concluding chapter.

The law of diminishing returns is one of the fundamental economic realities. Diminishing returns ultimately drives the existence of an optimal number of tornado or other hazard fatalities—the marginal cost of reducing tornado fatalities rises steeply after people exploit "easy" opportunities to reduce vulnerability. Eventually society reaches a point at which further reductions in (for example) tornado fatalities can actually lead to larger increases in early deaths elsewhere in society (Viscusi 1995). There definitely appears to be a hard core of fatalities that will be practically impossible to eliminate (e.g., permanent-home fatalities due to violent tornadoes). And our analysis of the life-saving effects of warnings found that the marginal benefit of a longer lead time is exhausted by 15 minutes; there are apparently no further casualty reductions to be realized from research to extend lead times beyond this. In the United States, at least, we may be approaching the "optimal" casualty level for some other types of hazardous weather. More than half of lightning fatalities, for example, occur with little warning on the periphery of thunderstorms (Lengyel and Brooks 2004), with many deaths now resulting from events that seem like "bolts out of the blue." The concept of balancing the cost (or tragedy if one prefers a more emotive term) of a fatality with the cost of avoiding fatalities is an important economic principle with application to all types of hazardous weather.

This book has examined the societal impacts of tornadoes, but some of the findings may translate to other types of extreme weather as well. One of our interesting findings concerns the value of tornado warnings. It would seem obvious that people should want to shelter when a tornado warning is issued for their area. And yet, when we applied some reasonable values to the parameters of this decision-making we found that responding did not always appear to be worthwhile, even though tornado warnings have become increasingly skillful over the past several decades. This raises the larger question concerning the value of other weather warnings. An expected utility framework has not been applied to ensure the value of many types of weather warnings.[4] The risk communication approach to hazard warnings implicitly assumes that following the recommended course of action is a worthwhile end, and views the public's unwillingness to respond as a product of a poorly designed or crafted message. A forecast or warning has low value when it does not allow the public to take an action to reduce; identify the factor that reduced the value of the warning: the size of the warned area (counties) relative to tornado damage paths. Once the source of low value is recognized, the warning can be revised accordingly, as with the reduction of the area of warnings.

NOTES

Note to Chapter 1

1. http://www.vortex2.org/home/

Notes to Chapter 2

1. See page 11 for description of F-scale.

2. The archive is available for download from the SPC Web site at: http://www.spc.noaa.gov/wcm/index.html#data.

3. Note that these totals will differ slightly from statistics reported for total tornadoes, which count multiple-state tornadoes as one entry as opposed to two or more tornadoes in our counts. The number of multistate tornadoes is small, however, and their inclusion as multiple-state segments will have no consequences of significance in our analysis.

4. Recall that these are state tornado segments.

5. Tornado Alley is defined as the states in the middle of the United States from Texas northward.

6. In no state was a statistically significant trend contrary to the national pattern observed, and in all the states that did not have a significant increase (decrease) in tornadoes (strong or violent tornadoes), there was no significant time trend.

7. The most-struck counties in any state have likely experienced a number of tornadoes between 1950 and 2007 above their true, long-run rates. Thus, it is hard to know if these frequently struck counties truly have high risk or if their rates are an artifact of less than 60 years of records.

8. A caveat is in order here, as very few counties in these states actually have population densities over 1,000 persons/mi^2.

9. Although not only the .05 significance level, compared to significance at better than the .01 level in Table 2.12.

Notes to Chapter 3

1. The population of the state employed for scaling is the mean state population reported in the decennial censuses between 1950 and 2000.

2. To illustrate, in a state with an average population of 1 million, the probability of observing zero fatalities in 58 years would be .5 if the annual tornado rate was .012. If the true fatality rate in the state were .012, it would be a 50-50 proposition to observe no fatalities over this period. Note that with the exception of California, the populations of the states with no fatalities were relatively small.

3. Tornadoes with a missing F-scale value are excluded from these calculations.

4. We generate this estimate by dividing the 43% of fatalities in mobile homes by the 7.6% of mobile home housing units and comparing it with the 31% of fatalities in permanent homes divided by either the 60% of units in the Census Bureau category 1, detached, which are single-family homes, or the 92.2% of units which would qualify as permanent homes.

5. The tornado records include the reported latitude and longitude of the points at which the tornado path began and ended, providing a way to map the path, and some hazards research has used this path information to map tornadoes into census tracts (Donner 2007). We plotted the paths of tornadoes into census tracts using GIS, but analysis revealed numerous inconsistencies with the counties that had been struck. For instance, of 241 tornadoes in Oklahoma in the 1990s that we could map into census tracts, the plotted tornado path was nowhere in the county or counties struck in 15% of the cases, meaning not a single census tract identified by the path was in the county or counties listed as struck by the tornado. Thus we concluded that counties were the only reliable measures we had for the tornado paths for a large enough sample for our statistical analysis.

6. On the general proposition of safety as a luxury good, see Viscusi, Vernon, and Harrington (2000, Chapter 19). Natural hazards papers that find a negative relationship between vulnerability and wealth include Anbarci et al. (2005), Kahn (2005), and Escaleras and Register (2008).

7. We would like to include mobile homes further back in the data set, but census documents do not report housing units by county before 1990.

8. The test is a chi-square test for a difference in proportions with five categories corresponding to the five parts of the day. The p-values in these tests are less than .001.

9. Hammer and Schmidlin (2002) document that many people fled the path of the May 3, 1999, Oklahoma City F5 tornado in their vehicles. Conceivably, mobile home residents might be less likely to leave in cars at night due to an inability to see and thus move away from an approaching tornado.

10. Of the states without dummy variables, Colorado has the highest number of tornadoes.

11. The effect of each included state variable on casualties relative to the omitted category was calculated, and the smallest of the state effects was normalized as equal to 1. The other state effects were divided by the effect of this smallest state.

12. The index was constructed based on the standard deviations of each of the four casualty measures. The score for each component was converted to the number of standard deviations above the lowest-ranked state, and the four casualty scores were then averaged to create the index.

13. Some narratives provided a number of "homes" damaged but did not specify a total for mobile homes, though these events had mobile-home fatalities. These narratives were excluded from the sample of buildings by type. If some mobile homes are reported as homes, then the probability for mobile homes will be overstated. We do not include different levels of damage in the analysis, because our calculations need to include all structures in the tornado path.

14. Sometimes, though, the number of persons at home will exceed the number of residents.

15. Residents of mobile homes may be relatively less likely to be at home when a tornado strikes if they receive a warning and abandon their mobile home, as recommended by the NWS. If this occurs to any large extent, the probability of death, as we have estimated it, would be biased downward from what we have estimated.

16. This limits our worst-case scenario because some states that have not experienced F5 tornadoes since 1900 will eventually experience an F5 tornado. Also some areas of large states like Texas that have experienced F5 tornadoes may not be vulnerable to tornadoes this strong. A state-level analysis seems a reasonable shortcut approximation of the area of potential F5 tornadoes.

17. These restrictions could be further refined, for instance, by considering the months of the year and the times of day that different states have experienced F5 tornadoes. But our sample of 65 F5 tornado segments hardly provides sufficient data to adequately explore the joint distributions of these different storm attributes.

18. For an extended critique of the Wurman et al. projections, see Brooks, Doswell, and Sutter (2008).

Notes to Chapter 4

1. Of course, there are certain people who try to take pictures of the tornado instead of taking cover.

2. Katz and Murphy (1997) provide an analysis of the value of a forecast from a meteorological perspective, including the more complicated case of a continuous variable forecast. Note that although tornado warnings are not technically probabilistic forecasts, the error probabilities still matter for value determination, and so any imperfect forecast or warning is inherently probabilistic.

3. Note that for the POD, FAR, and other probabilities discussed here, household response depends on residents' subjective perception of the probability, and not the "true" or best expert estimate. We will not explicitly state this in discussing each component of the decision, but misperception can have the same effect as a change in the actual probability.

4. Warning response in a community will not then have a single threshold, but response could be relatively small due to a recent false alarm.

5. The NWS false alarm ratio is based on tornado warnings issued for each county and may deviate from common perceptions of false alarms. For example, if warnings are issued for three counties and a tornado strikes only one county, the FAR will be .666, even though many people might view the warnings as a group and would not consider this a false alarm. For other critiques of the NWS definition of false alarms, see Drobot et al. (2007).

6. A recent study of FEMA mitigation projects (Multihazard Mitigation Council 2005) provides an alternative means to value tornado injuries. This study assumes tornado casualties are geometrically distributed across a five-point severity scale, with monetized values of injuries ranging from $6,000 to $2.4 million, and yields a value of a statistical injury for tornadoes of about $40,000. Given that injuries account for about 2% of the monetary impact of tornadoes, the cost per injury would need to substantially exceed the $76,000 figure to materially affect the total.

7. Reported at http://ftp.bls.gov/pub/suppl/empsit.cseeb2.txt.

8. For an extended discussion of the construction of this variable, refer to Simmons and Sutter (2005).

9. The NWS tracked two dates for radar installation at WFOs, the installation date and a commissioning date, which was a more formal ceremony. The radars were operational as of the installation date, so this is the proper date to use in constructing our treatment variable. Initially, however, we were supplied with the commissioning dates, and we constructed a treatment variable based on these dates. This alternative Doppler radar variable had a point estimate for fatalities and injuries of about half of the magnitude as the variable based on installation dates, and the point estimate failed to attain significance. This is exactly the effect we would expect with such a misclassified treatment variable, as some tornadoes that occur with Doppler radar in

place and with reduced casualties would be classified as occurring without Doppler based on commissioning dates.

10. Note that since we are using the warning for the first county of a tornado path, the probability of detection and false alarm ratio figures cited in this and the following section will not correspond with official NWS verification statistics, based on all county warnings.

11. We apply current CWA through our sample, even though WFOs were reorganized with the modernization of the NWS in the 1990s. The counties contained in each Nielsen Designated Market Area are reported in the annual editions of the *Broadcasting and Cable Yearbook*.

12. A complication arises because the FAR is undefined when tornadoes occur in a geography where no warnings have been issued over the prior one or two years. It is unclear how residents will treat warnings in such instances. Most of these tornadoes occur in states with low tornado frequency. We control for these cases by setting the FAR equal to 1 and including a dummy variable to indicate the undefined FAR. Exclusion of these tornadoes from the estimation does not affect our results in a substantive manner.

13. The median here is the median of lead times for tornadoes occurring within tornado watches.

14. 99.1% of tornadoes in 1996–1997 occurred in counties where WSR-88D radar had been installed. Radar installation was completed by the end of 1997.

15. Reported at http://ftp.bls.gov/pub/suppl/empsit.cseeb2.txt.

16. The overall response rate and the portion of the population within the SBW polygon place a lower bound on the response rate outside of the polygon. For instance, if the overall response rate was 50%, then even if the response rate in the polygon was 100%, the response rate outside of the polygon would have to be almost 30%.

17. For more details on these calculations, as well as a sensitivity analysis of how the estimated savings vary with each of the parameters of the calculation, see Sutter and Erickson (2010).

Notes to Chapter 5

1. See Richard Monastersky, "Shelter in the Storm: Oklahoma Tornadoes Give 'Strong Rooms' Their First Test," *Science News*, Volume 155, Number 21, p.335, May 1999.

2. Details from the *Peoria Journal Star*, July 15, 2004.

3. A tornado represents a correlated risk of death or injury for family members, creating the likelihood that two or more members might be killed or injured at once. This might be considered a markedly worse outcome than independent risks.

4. In the long run, and assuming that the underlying national proportion of fatalities in mobile homes did not change.

5. In the May 3, 1999, Oklahoma tornadoes, 79% of injuries occurred in permanent homes, consistent with this speculation. But 63% of fatalities in this tornado outbreak also occurred in permanent homes, so we simply lack the data to estimate the distribution of injuries with any precision.

6. Note that the rank order of cost per fatality or casualty avoided in each type of home is based on the rank orders of the probabilities of tornado damage in Table 2.7.

7. One elderly resident died in a fall trying to get into an underground shelter during the May 3, 1999, tornadoes in Oklahoma. (See the *Storm Events* narrative on page 102 for details.) The monetized value of fatalities prevented in permanent homes was $600 over the life of a shelter in Mississippi, and if sheltering casualties are no more than 10% of direct tornado casualties, the value will be $60 or less in all states.

Notes to Chapter 6

1. Grazulis (1993, 1997) reports damage estimates for only 19 of the 29 tornadoes prior to 1995 with reported damage over $50 million in the SPC archive.

2. We also estimated the models with the time of day and month of year variables included in the casualties analysis, even though we would expect that timing should not affect damage. The variables failed to attain significance jointly, and had no notable impact on the inferences regarding other variables. The insignificance of the timing variables in damage analysis implies that the casualties results are likely due to differences in vulnerability of residents, and not variations in the severity of the tornado not captured by the F-scale or path-length variables.

3. For studies of the economic impact of utility lifeline disruptions, see Rose et al. (1997) and Rose and Lim (2002).

4. For more on labor market impacts of hurricanes, see also Ewing and Kruse (2002) and Ewing, Kruse, and Thompson (2005).

Notes to Chapter 7

1. Damage totals are from SPC (n.d.) and adjusted for inflation using the CPI for housing (series CUUR0000JAA), available at http://data.bls.gov/cgi-bin/surveymost?cu.

2. For details on the estimated time savings with SBWs, including a sensitivity analysis, see Sutter and Erickson (2010).

3. And no offsetting increases in fatalities in other tornadoes.

4. Expected utility provides a way to compare benefits and costs of a given action.

BIBLIOGRAPHY

Chapter 1

Chapman, R. E., 1992: *Benefit-Cost Analysis for the Modernization and Associated Restructuring of the National Weather Service*. Silver Spring MD: U.S. Department of Commerce.

Crum, T. D., R. E. Saffle, and J. W. Wilson, 1998: An Update on the NEXRAD Program and Future WSR-88D Support to Operations. *Weather and Forecasting*, **13**, 253–262.

Friday Jr., E. W., 1994: The Modernization and Associated Restructuring of the National Weather Service: An Overview. *Bulletin of the American Meteorological Society*, **75**, 43–52.

Simmons, K. M., and D. Sutter, 2005: WSR-88D Radar, Tornado Warnings, and Tornado Casualties. *Wea. Forecasting*, **20**, 301–310.

Chapter 2

Anderson, C. J., C. K. Wikle, Q. Zhou, and J. A. Royle, 2005: Population Influences on Tornado Reports in the United States. Unpublished manuscript, Iowa State University.

Beamish, J. G., R. C. Goss, J. H. Aitles, and Y. Kim, 2001: Not a trailer anymore: Perceptions of Manufactured Housing. *Housing Policy Debate*, **12**, 373–392

Doswell III, C. A., and D. W. Burgess, 1988: On Some Issues of United States Climatology. *Monthly Weather Review*, **116**, 495–501.

Multihazard Mitigation Council, 2005: Natural Hazard Mitigation Saves Lives: An Independent Study to Assess the Future Savings from Mitigation Activities. 2 vols, National Institute of Building Sciences, Washington, DC.

Ray, P. S., P. Bieringer, X. Niu, and B. Whissel, 2003: An Improved Estimate of Tornado Occurrence in the Central Plains of the United States. *Monthly Weather Review*, **131**, 1026–1031.

Schaefer, J. T., R. S. Schneider, and M. P. Kay, 2002: The Robustness of Tornado Hazard Estimates. Preprints, *Third Symposium on Environmental Applications*, 13–17, January 2002, 35–41.

Schaefer, J. T., D. L. Kelly, and R. F. Abbey, 1986: A Minimum Assumption Tornado-Hazard Probability Model. *J. Cli. Appl. Met.*, **25**, 1934–1945.

Texas Tech Fujita Scale Report 2006. A Recommendation for an Enhanced Fujita Scale. Report from the Wind Science and Engineering Center. Texas Tech University.

Chapter 3

Aguirre, Begnino E., 1988: The Lack of Warnings Before the Saragosa Tornado. *International Journal of Mass Emergencies and Disaster*, **6**, 65–74.

Anbarci, N., M. Escaleras, and C. A. Register, 2005: Earthquake Fatalities: The Interaction of Nature and Political Economy. *Journal of Public Economics*, **89**, 1907–1933.

Ashley, W. S., 2007: Spatial and Temporal Analysis of Tornado Fatalities in the United States, 1880–2005. *Wea. Forecasting*, **22**, 1214–1228.

Ashley, W. S., A. J. Krmenec, and R. Schwantes, 2008: Nocturnal Tornadoes. *Weather and Forecasting*, **23**, 795–807.

Boruff, Bryan J., Jaime A. Easoz, Steve D. Jones, Heather R. Landry, Jaime D. Mitchem, and Susan L. Cutter, 2003: Tornado Hazards in the United States. *Climate Research*, **24**, 103–117.

Brooks, H. E., and C. A. Doswell III, 2002: Deaths in the 3 May 1999 Oklahoma City Tornado from a Historical Perspective. *Weather and Forecasting*, **17**, 354–361.

Brooks, H. E., C. A. Doswell III, and D. Sutter, 2008: Comments on "Low Level Winds in Tornadoes and Potential Catastrophic Tornado Impacts in Urban Areas." *Bulletin of the American Meteorological Society*, 89(1):87–90, 2008.

Brown, S., P. Archer, E. Kruger, and S. Mallonee, 2002: Tornado-Related Deaths and Injuries in Oklahoma due to the 3 May 1999 Tornadoes. *Wea. Forecasting*, **17**, 343–353.

Census Bureau 2000 (available online at www.census.gov).

Donner, W. R., 2007: The Political Economy of Disaster: An Analysis of Factors Influencing U.S. Tornado Fatalities and Injuries, 1998–2000. *Demography*, **44**, 669–685.

Doswell III, C. A., and D. W. Burgess, 1988: On Some Issues of United States Climatology. *Monthly Weather Review*, **116**, 495–501.

Escaleras, M., and C. A. Register, 2008: Mitigating Natural Disasters through Collective Action: The Effectiveness of Tsunami Early Warnings, *Southern Economic Journal*, 74(4): 1017–1034.

Golden, J. H., and C. R. Adams, 2000: The Tornado Problem: Forecast, Warning, and Response. *Natural Hazards Review*, **1**, 107–118.

Golden, J. H., and Snow, J. T., 1991: Mitigation Against Extreme Windstorms. *Review of Geophysics*, **29**, 477–504.

Grazulis, T. P., 1993: Significant Tornadoes 1680–1991. Environmental Films, St. Johnsbury, Vt.

Hammer, B., and T. W. Schmidlin, 2002: Response to Warnings during the 3 May 1999 Oklahoma City Tornado: Reasons and Relative Injury Rates. *Weather and Forecasting*, **17**, 577–581.

Kahn, M. E., 2005: The Death Toll From Natural Disasters: The Role of Income, Geography, and Institutions. *Review of Economics and Statistics*, **87**, 271–284.

Multihazard Mitigation Council, 2005: Natural Hazard Mitigation Saves Lives: An Independent Study to Assess the Future Savings from Mitigation Activities. 2 vols, National Institute of Building Sciences, Washington, DC.

Simmons, K. M., and D. Sutter, 2005a: Protection From Nature's Fury: An Analysis of Fatalities and Injuries from F5 Tornadoes. *Natural Hazards Review*, **6**, 82–87.

Simmons, K. M., and D. Sutter, 2005b: WSR-88D Radar, Tornado Warnings, and Tornado Casualties. *Weather and Forecasting*, **20**, 301–310.

Simmons, K. M., and D. Sutter, 2008: Tornado Warnings, Lead Times and Tornado Casualties: An Empirical Investigation. *Wea. Forecasting*, **23**, 246–258.

Storm Events National Climatic Data Center (NCDC) (available online at http://www4.ncdc.noaa.gov/cgi-win/wwcgi.dll?wwEvent~Storms).

Viscusi, W. K., J. M. Vernon, and J. E. Harrington, Jr., 2000: *Economics of Regulation and Antitrust*, 3rd Edition. Cambridge MA: MIT Press.

Wurman, J., C. Alexander, P. Robinson, and Y. Richardson, 2007: Low-level winds in tornadoes and potential catastrophic tornado impacts in urban areas. *Bull. Amer. Meteorol. Soc.*, **88**, 31–46.

Chapter 3 Appendix

American Community Survey 2006 (available online at http://www.census.gov/acs/www/Products/).

Census Bureau 1990 (available online at www.census.gov).

Census Bureau 2000 (available online at www.census.gov).

Chapter 4

Aguirre, B. E., 1988: The Lack of Warnings Before the Saragosa Tornado. *International Journal of Mass Emergencies and Disaster*, **6**, 65–74.

Ashley, W. S., 2007: Spatial and Temporal Analysis of Tornado Fatalities in the United States, 1880–2005. *Wea. Forecasting*, **22**, 1214–1228.

Balluz, L., T. Holmes, J. Malilay, L. Schieve, and S. Kiezak, 2000: Predictors for People's Response to a Tornado Warning: Arkansas, 1 March 1997. *Disasters*, **24**, 71–77.

Barnes, L. R., E. C. Gruntfest, M. H. Hayden, D. M. Schultz, and C. Benight, 2007: False Alarms and Close Calls: A conceptual model of warning accuracy. *Wea. Forecasting*, **22**, 1140–1147.

Bieringer, P., and P. S. Ray, 1994: A Comparison of Tornado Warning Lead Times with and without NEXRAD Doppler Radar. *Weather and Forecasting*, **11**, 47–52.

Bureau of Labor Standards 2007 (available online at http://ftp.bls.gov/pub/suppl/ empsit.cseeb2.txt.)

Broadcasting and Cable Yearbook 2009. New Providence NJ: Bowker Press.

Brooks, H. E., 2004: Tornado-Warning Performance in the Past and Future: A Perspective from Signal Detection Theory. *Bull. Amer. Meteor. Soc.*, **85**, 837–843.

Brown, S., P. Archer, E. Kruger, and S. Mallonee, 2002: Tornado-Related Deaths and Injuries in Oklahoma due to the 3 May 1999 Tornadoes. *Weather and Forecasting*, **17**, 343–353.

Carter, A. O., M. E. Millson, and D. E. Allen, 1989: Epidemiologic Study of Deaths and Injuries Due to Tornadoes. *American Journal of Epidemiology*, **130**, 1209–1218.

Cesario, F. J., 1976: Value of Time in Recreation Benefit Studies. *Land Economics*, **52**, 32–41.

Chapman, R. E., 1992: *Benefit-Cost Analysis for the Modernization and Associated Restructuring of the National Weather Service*. Silver Spring MD: U.S. Department of Commerce.

Crum, T.D., R. E. Saffle, and J. W. Wilson, 1998: An Update on the NEXRAD Program and Future WSR-88D Support to Operations. *Weather and Forecasting*, **13**, 253–262.

Dow, K., and S. L. Cutter, 1998: Crying Wolf: Repeat Responses to Hurricane Evacuation Orders. *Coastal Management*, **26**, 237–252.

U.S. Environmental Protection Agency, 1997: *The Benefits and Costs of the Clean Air Act, 1970 to 1990*. [Available online at http://www.epa.gov/airprogm/oar/sect812/ index.html]

Hammer, B., and T. W. Schmidlin, 2002: Response to Warnings during the 3 May 1999 Oklahoma City Tornado: Reasons and Relative Injury Rates. *Wea. Forecasting*, **17**, 577–581.

Hodler, T. W., 1982: Residents' Preparedness and Response to the Kalamzoo Tornado. *Disasters*, **6**, 44–49.

Jacks, Eli, and John Ferree, 2007: Socio-Economic Impacts of Storm-Based Warnings. Paper presented at the 2nd Symposium in Policy and Socio-Economic Impacts, San Antonio, TX, January.

Katz, R. W., and A. H. Murphy (eds), 1997: *Economic Value of Weather and Climate Forecasts*. Cambridge: Cambridge University Press.

Laffont, J. J., 1989: *The Economics of Uncertainty and Information*. Cambridge MA: MIT Press.

Legates, D. R., and M. D. Biddle, 1999: Warning Response and Risk Behavior in the Oak Grove - Birmingham, Alabama, Tornado of 8 April 1998. Quick Response Report #116, Natural Hazards Research and Applications Information Center, University of Colorado.

Letson, D., D. Sutter and J. Lazo, 2007: The Economic Value of Hurricane Forecasts. *Natural Hazards Review* 8(3): 78–86.

Liu, S., L. E. Quenemoen, J. Malilay, E. Noji, T. Sinks, and J. Mendlein, 1996: Assessment of a Severe-Weather Warning System and Disaster Preparedness, Calhoun County Alabama 1994. *American Journal of Public Health*, **86**, 87–89.

Looney, Michael, 2006: Polygon Warnings: The Sharp Focus on Service. Presentation to the Central Regional Managers' Conference, Kansas City, March 7.

Multihazard Mitigation Council, 2005: Natural Hazard Mitigation Saves Lives: An Independent Study to Assess the Future Savings from Mitigation Activities. 2 vols, National Institute of Building Sciences, Washington, DC.

Paul, B. K., V. T. Brock, S. Csiki, and L. Emerson, 2003: Public Response to Tornado Warnings: A Comparative Study of the May 4, 2003, Tornados in Kansas, Missouri and Tennessee. Quick Response Research Report #165, Natural Hazards Research and Applications Information Center, University of Colorado.

Schmidlin, T. W., B. O. Hammer, Y. Ono, and P. S. King, 2009: Tornado Shelter-Seeking Behavior and Tornado Shelter Options Among Mobile Home Residents in the United States. *Nat. Hazards*, **48**, 191–201.

Simmons, K. M. and D. Sutter, 2009: False Alarms, Tornado Warnings and Tornado Casualties. *Weather, Climate and Society*, **1**, 38–53.

Smith, V. K., W. H. Desvousges, and M. P. McGivney, 1983: The Opportunity Cost of Travel Time in Recreation Demand Models. *Land Economics*, **59**, 259–278.

Sorensen, J. H., 2000: Hazard Warning Systems: Review of 20 Years of Progress. *Nat. Hazards Rev.*, **1**, 119–125.

Storm Prediction Center Historical Severe Storm Database (available online at http://www.spc.noaa.gov/wcm/index.html#data).

Sutter, D., and S. Erickson, 2010: The Value of Tornado Warnings and Improvements in Warnings. *Weather, Climate and Society*, **2**, 103–112

Tiefenbacher, J. P., W. Monfredo, M. Shuey, and R. J. Cecora, 2001: Examining a 'Near-Miss' Experience: Awareness, Behavior, and Post-Disaster Response Among Residents on the Periphery of a Tornado-Damage Path. Quick Response Research Report #137, Natural Hazards Research and Applications Information Center, University of Colorado.

Viscusi, W. K., 1993: The Value of Risks to Life and Health. *Journal of Economic Literature*, **31**, 1912–1946.

Viscusi, W. K., 2004: The Value of Life: Estimates with Risks by Occupation and Industry. *Economic Inquiry*, **42**, 29–48.

Viscusi, W. K., and Joseph E. Aldy. 2003: The Value of a Statistical Life: A Critical Review of Market Estimates Throughout the World. *Journal of Risk and Uncertainty*, 27(1): 5–76.

Viscusi, W. K., J. M. Vernon, and J. E. Harrington, Jr., 2000: *Economics of Regulation and Antitrust*, 3rd Edition. Cambridge MA: MIT Press.

Chapter 5

Arrow, K. A., and A. C. Fisher, 1974: Environmental Preservation, Uncertainty, and Irreversibility. *Quarterly Journal of Economics*, **88**, 312–319.

Arrow, K. A., and R. C. Lind, 1970: Uncertainty and the Evaluation of Public Investment Decisions. *American Economic Review*, **60**, 364–378.

Beron, K. J., J. C. Murdoch, M. A. Thayer, and W. P. M. Vijverberg, 1997: An Analysis of the Housing Market Before and After the 1989 Loma Prieta Earthquake. *Land Economics*, **73**, 101–113.

Brown, S., P. Archer, E. Kruger, and S. Mallonee, 2002: Tornado-Related Deaths and Injuries in Oklahoma due to the 3 May 1999 Tornadoes. *Weather and Forecasting*, 17:343–353.

Camerer, Colin F., and Howard Kunreuther, 1989: Decision Processes for Low Probability Events: Policy Implications. *Journal of Policy Analysis and Management*, 8(4): 565–92.

Carter, A. O., M. E. Millson, and D. E. Allen, 1989: Epidemiologic Study of Deaths and Injuries Due to Tornadoes. *American Journal of Epidemiology*, 130: 1209–1218.

Ewing, B. T., and J. B Kruse, 2006: Valuing self protection: income and certification effects for safe rooms. *Construction Management and Economics*, 24(10), 1057–68.

Federal Emergency Management Agency, 1999: Taking Shelter From the Storm: *Building a Safe Room Inside Your House*, 2nd edition. Publication #320, Washington DC.

Federal Emergency Management Agency, 2000: *Design and Construction Guidance for Community Shelters*. Publication #361, Washington DC.

Kahneman, D., and A. Tversky, 1979: Prospect Theory: An Analysis of Decision Under Risk, *Econometrica*, **47**, 263–291.

Kunreuther, H., 1978: *Disaster Insurance Protection: Public Policy Lessons*. New York: John Wiley.

Kunreuther, H., and M. Pauly, 2004: Neglecting Disaster: Why Don't People Insure Against Large Losses? *Journal of Risk and Uncertainty*, **28**, 5–21.

Meyer, Robert J., 2006: Why We Under-Prepare for Hazards. In *On Risk and Disaster: Lessons From Hurricane Katrina*, edited by R.J. Daniels, D.F. Kettl, and H. Kunreuther, pp. 153–173. Philadelphia: University of Pennsylvania Press.

Merrell, D., K. M. Simmons, and D. Sutter, 2005: The Determinants of Tornado Casualties and the Benefits of Tornado Shelters. *Land Economics*, **81**, 87–99.

Monastersky, R., 1999: Shelter in the Storm: Oklahoma Tornadoes Give 'Strong Rooms' Their First Test. *Science News*, **155**, May 22, 1999, p. 335.

Oklahoma Manufactured Housing Association, 2005: Manufactured home park report. Oklahoma City, OK.

Ozdemir, Ozlem, 2005: Risk Perception and the Value of Safe Rooms as a Protective Measure from Tornadoes: A Survey Method. In B. T. Ewing and J. B. Kruse (eds.) *Economics and Wind*. Nova Science Publishing, pp. 89–104.

Palm, R., 1998: The Demand for Disaster Insurance: Residential Coverage. In *Paying the Price*, edited by H. Kunreuther and R. J. Roth, Sr., pp. 51–66. Washington DC: Joseph Henry Press.

Peoria Journal Star

Schmidlin, T. W., B. O. Hammer, Y. Ono, and P. S. King, 2009: Tornado Shelter-Seeking Behavior and Tornado Shelter Options Among Mobile Home Residents in the United States. *Nat. Hazards*, **48**, 191–201.

Schmidlin, T. W., B. Hammer, and J. Knabe, 2001: Tornado Shelters in Mobile Home Parks in the United States. *Journal of the American Society of Professional Emergency Planners*, **8**, 1–15.

Simmons, K. M., J. B. Kruse, and D. A. Smith, 2002: Valuing mitigation: real estate market response to hurricane loss reduction measures. *Southern Economic Journal*, **68**, 660–671.

Simmons, K. M., and D. Sutter, 2006: Direct estimation of the cost effectiveness of tornado shelters. *Risk Analysis*, **26**, 945–954.

Simmons, K. M., and D. Sutter, 2007: Tornado Shelters and the Housing Market. *Construction Management and Economics*, **25**, 1117–1124.

Simmons, K. M., and D. Sutter, 2007: Tornado Shelters and the Manufactured Home Parks Market. *Natural Hazards*, **43**, 365–378.

Spreyer, J. F., and W. R. Ragas, 1991: Housing prices and flood risk: an examination using spline regression. *Journal of Real Estate Finance and Economics*, **4**, 395–407.

Sutter, D., and M. Poitras, 2010: Do People Respond to Low Probability Risks? Evidence from tornado risk and manufactured homes. *Journal of Risk and Uncertainty*, **40**, 181–196.

Sutter, D., and E. Stephenson, 2008: Political Economy and Natural Hazards Mitigation: State Incentives for Tornado Shelters. *Journal of Public Finance and Public Choice*, **25**, 77–92.

Viscusi, W. K., J. M. Vernon, and J. E. Harrington, Jr., 2000: *Economics of Regulation and Antitrust*, 3rd Edition. Cambridge MA: MIT Press.

Samuelson, W., and R. Zeckhauser, 1988: Status Quo Bias in Decision Making. *Journal of Risk and Uncertainty*, **1**, 7–59.

Chapter 6

Belasen, A., and S. W. Polachek, 2009: How Disasters Affect Local Labor Markets: The Effects of Hurricanes in Florida. *Journal of Human Resources*, **44**, 251–276.

Brooks, Harold E., and Charles A. Doswell, III., 2001: Normalized Damage from Major Tornadoes in the United States: 1890–1999. *Weather and Forecasting*, 16(1): 168–176.

Ewing, B., and J. B. Kruse, 2002: The Impact of Project Impact on the Wilmington, NC Labor Market. *Public Finance Review*, **30**, 296–309.

Ewing, B., J. B. Kruse, and M. Thompson, 2003: A Comparison of Employment Growth and Stability Before and After the Fort Worth Tornado. *Environmental Hazards*, **5**, 83–91.

Ewing, B., J. B. Kruse, and M. Thompson, 2005: Tornadoes and Labor Markets: Fort Worth, Nashville, and Oklahoma City. *Economics and Wind*. New York: Nova Science Press.

Ewing, B., J. B. Kruse, and M. Thompson, 2005: An Empirical Examination of the Corpus Christi Unemployment Rate and Hurricane Bret. *Natural Hazards Review*, **4**, 191–196.

Ewing, Bradley, Jamie B. Kruse, and Mark A. Thompson, 2007: Twister! Employment Responses to the 3 May 1999 Oklahoma City Tornado. *Applied Economics*. iFirst, 1–12.

Ewing, B., J. B. Kruse, and Y. Wang, 2007: Local Housing Price Index Analysis in Wind-Disaster-Prone Areas, *Natural Hazards*, **40**, 463–483.

Gall, M., K. A. Borden, and S. L. Cutter, 2009: When Do Losses Count? Six Fallacies of Natural Hazard Loss Data. *Bull. Amer. Meteor. Soc.*, **90**, 799–809.

Grazulis, T. P., 1993: Significant Tornadoes 1680–1991. Environmental Films, St. Johnsbury, VT.

Grazulis, T. P., 1997: Significant Tornadoes, Update 1992–1995. Environmental Films, St. Johnsbury, VT.

Greene, W. H., 2000: *Econometric Analysis*, 4th ed. Prentice Hall.

Guimares, P., F. Hefner, and D. Woodward, 1993: Wealth and Income Effects of Natural Disasters: An Econometric Analysis of Hurricane Hugo. *Review of Regional Studies*, **23**, 97–114.

Heckman, J. J., 1976: The Common Structure of Statistical Models of Truncation, Sample Selection, and Limited Dependent Variables and a Simple Estimator for Such Models. *Annals of Economic and Social Measurement*, **5**, 475–492.

Institute for Business and Home Safety (available online at http://www.disastersafety.org/text.asp?id=commlines).

Martinez, M., and B. Ewing, 2008: A Look at the Economic Impact of Tornado Induced Damage in Tulia, Texas. Paper presented at 2008 Hazards and Disaster Researchers Meeting, Broomfield, CO.

Mill, J., 1848: *Principles of Political Economy*. New York: A M Kelley Publishers.

Pielke Jr., R. A., J. Gratz, C. W. Landsea, D. Collins, M. Saunders, and R. Musulin, 2008: Normalized Hurricane Damage in the United States: 1900–2005. *Natural Hazards Review*, 29–42.

Pielke, Jr., R. A., and C. W. Landsea, 2005: Normalized Hurricane Damages in the United States: 1925–1995. *Weather and Forecasting*, **13**, 621–31,

Rose et al 1997

Rose, A., and D. Lim, 2002: Business Interruption Losses from Natural Hazards: Conceptual and Methodological Issues in the Case of the Northridge Earthquake. *Environmental Hazards*, **4**, 1–14.

Smith, V. K., J. C. Carbone, J. C. Pope, D. G. Hallstrom, and M. E. Darden, 2006: Adjusting to Natural Disasters. *Journal of Risk and Uncertainty*, **33**, 37–54.

Stogsdill, S., 2009: Picher projects its end as official municipality. *Tulsa World*, June 23.

Tierney, K. J., 1997: Business Impacts of the Northridge Earthquake. *Journal of Contingencies and Crisis Management*, **5**, 87–97.

Webb, G. R., K. J. Tierney, and J. M. Dalhamer, 2000: Businesses and Disasters: Empirical Patterns and Unanswered Questions. *Natural Hazards Review*, **1**, 83–90.

Webb, Gary R., Kathleen J. Tierney, and James M. Dalhamer, 2003: Predicting Long-Term Business Recovery from Disaster: A Comparison of the Loma Prieta Earthquake and Hurricane Andrew. *Environmental Hazards*, **4**, 45–58.

West, C. T., and D. G. Lenze, 1994: Modeling the Regional Impact of Natural Disaster and Recovery: A General Framework and an Application to Hurricane Andrew. *International Regional Science Review*, **17**, 121–150.

Wurman, J., C. Alexander, P. Robinson, and Y. Richardson, 2007: Low-level winds in tornadoes and potential catastrophic tornado impacts in urban areas. *Bull. Amer. Meteorol. Soc.*, **88**, 31–46.

Yoshida, K., and R. E. Deyle, 2005: Determinants of Small Business Hazard Mitigation. *Natural Hazards Review*, **6**, 1–12.

Chapter 7

Bureau of Labor Standards 2007 (available online at http://ftp.bls.gov/pub/suppl/empsit.cseeb2.txt).

Brooks, H. E., and C. A. Doswell III, 2002: Deaths in the 3 May 1999 Oklahoma City Tornado from a Historical Perspective. *Weather and Forecasting*, **17**, 354–361.

De Alessi, L., 1996: Error and Bias in Benefit-Cost Analysis: HUD's Case for the Wind Rule. *Cato Journal*, **16**, 129–147.

Doswell, C. A., A. R. Moeller, and Harold E. Brooks, 1999: Storm Spotting and Public Awareness Since the First Tornado Forecasts of 1948. *Weather and Forecasting*, **14**, 544–557.

Golden, J. H., and C. R. Adams, 2000: The Tornado Problem: Forecast, Warning, and Response. *Natural Hazards Review*, **1**, 107–118.

Grazulis, T. P., 1993: Significant Tornadoes 1680–1991. Environmental Films, St. Johnsbury, Vt.

Grosskopf, K. R., 2005: Assessing the Effectiveness of Mitigation: A Case Study of Manufactured Housing and the 2004 Hurricane Season. *Journal of Emergency Management*, **3**, 27–32.

Hammer, B., and T. W. Schmidlin, 2002: Response to Warnings during the 3 May 1999 Oklahoma City Tornado: Reasons and Relative Injury Rates. *Wea. Forecasting*, **17**, 577–581.

Lengyel, M. M., and H. E. Brooks, 2004: Lightning Casualties and their Proximity to Surrounding Cloud-to-Ground Lightning. Unpublished manuscript, National Severe Storms Laboratory.

National Academy of Sciences, 2002: *Weather Radar Technology Beyond NEXRAD*. National Academies Press (available online at http://www.nap.edu/openbook/0309084660/html/1.html).

Oklahoma Manufactured Housing Association, 2005: Manufactured home park report. Oklahoma City, OK.

Rappaport, E., 2000: Loss of Life in the United States Associated with Recent Atlantic Tropical Cyclones. *Bulletin of the American Meteorological Society*, **81**, 2065–2073.

Schmidlin, T. W., B. O. Hammer, Y. Ono, and P. S. King, 2009: Tornado Shelter-Seeking Behavior and Tornado Shelter Options Among Mobile Home Residents in the United States. *Nat. Hazards*, **48**, 191–201.

Schmidlin, T. W., B. Hammer, and J. Knabe, 2001: Tornado Shelters in Mobile Home Parks in the United States. *Journal of the American Society of Professional Emergency Planners*, **8**, 1–15.

Simmons, K. M., and D. Sutter, 2008: Manufactured Home Building Regulations and the February 2, 2007 Florida Tornadoes. *Natural Hazards*, **46**, 415–425.

Sutter, D., and S. Erickson, 2010: The Value of Tornado Warnings and Improvements in Warnings. *Weather, Climate and Society*, **2**, 103–112

Vescio, M. D., and R. L. Thompson, 2001: Subjective Tornado Probability Forecasts in Severe Weather Watches. *Wea. Forecasting*, **16**, 192–195.

Viscusi, W. K., 1994: Mortality Effects of Regulatory Costs and Policy Evaluation Criteria. *Rand Journal of Economics*, **25**, 94–109.

Wurman, J., C. Alexander, P. Robinson, and Y. Richardson, 2007: Low-level winds in tornadoes and potential catastrophic tornado impacts in urban areas. *Bull. Amer. Meteorol. Soc.*, **88**, 31–46.

INDEX